International Conference on Architecture and Civil Engineering (ICACE 2017)

AA004680

IOP Conference Series: Materials Science and Engineering
Volume 291

Petaling Jaya, Malaysia
8 - 9 May 2017

Editor:

Nordin Abd Rahman

ISBN: 978-1-5108-5503-8
ISSN: 1757-8981

Printed from e-media with permission by:

Curran Associates, Inc.
57 Morehouse Lane
Red Hook, NY 12571

Some format issues inherent in the e-media version may also appear in this print version.

Copyright© (2017) by the Institute of Physics
All rights reserved. The material featured in this book is subject to
IOP copyright protection, unless otherwise indicated.

Printed by Curran Associates, Inc. (2018)

For permission requests, please contact the Institute of Physics
at the address below.

Institute of Physics
Dirac House, Temple Back
Bristol BS1 6BE UK

Phone: 44 1 17 929 7481
Fax: 44 1 17 920 0979

techtracking@iop.org

Additional copies of this publication are available from:

Curran Associates, Inc.
57 Morehouse Lane
Red Hook, NY 12571 USA
Phone: 845-758-0400
Fax: 845-758-2633
Email: curran@proceedings.com
Web: www.proceedings.com

Table of contents

Volume 291

International Conference on Architecture and Civil Engineering (ICACE 2017)
8–9 May 2017, Petaling Jaya, Malaysia

Accepted papers received: 21 December 2017
Published online: 2 January 2018

Preface

Papers

IOP Conf. Series: Materials Science and Engineering **291** (2017) 011001 doi:10.1088/1757-899X/291/1/011001

Preface:

This volume presents a compilation of research works covering the fields of architecture and civil engineering. All manuscripts in this volume were presented during the First International Conference on Architecture and Civil Engineering 2017 which was held at Armada Hotel, Petaling Jaya, Malaysia on 8-9 May 2017. The editor(s) of the proceeding would like to express the utmost gratitude and thanks to all reviewers in the technical team for making this volume a success.

Editor:

Dr. Nordin Abd Rahman

Content from this work may be used under the terms of the Creative Commons Attribution 3.0 licence. Any further distribution of this work must maintain attribution to the author(s) and the title of the work, journal citation and DOI.
Published under licence by IOP Publishing Ltd

IOP Conf. Series: Materials Science and Engineering **291** (2017) 011001 doi:10.1088/1757-899X/291/1/011001

IOP Conf. Series: Materials Science and Engineering **291** (2017) 011002 doi:10.1088/1757-899X/291/1/011002

Peer review statement

All papers published in this volume of *IOP Conference Series: Materials Science and Engineering* have been peer reviewed through processes administered by the proceedings Editors. Reviews were conducted by expert referees to the professional and scientific standards expected of a proceedings journal published by IOP Publishing.

Content from this work may be used under the terms of the Creative Commons Attribution 3.0 licence. Any further distribution of this work must maintain attribution to the author(s) and the title of the work, journal citation and DOI.
Published under licence by IOP Publishing Ltd

Criteria for Public Open Space Enhancement to Achieve Social Interaction: a Review Paper

S A Salih[1] and S Ismail[2]

[1] Master Student at Department of Architecture, Faculty of Design and Architecture, Universiti Putra Malaysia, Jalan UPM, 43400 Serdang, Selangor, Malaysia.

[2] Lecturer at Department of Architecture, Faculty of Design and Architecture, Universiti Putra Malaysia, Jalan UPM, 43400 Serdang, Selangor, Malaysia.

E-mail Address: sarah_alsaadi1990@yaho.com, sumarni.upm@gmail.com.

Abstract.
A This paper presents a various literatures, studies, transcripts and papers aiming to provide an overview of some theories and existing research on the significance of natural environments and green open spaces to achieve social interaction and outdoor recreation. The main objective of the paper is to identify the factors that affecting social interaction in green open spaces, through proving that an appropriate open spaces is important to enhance social interaction and community. This study employs (qualitative) summarizing content analysis method which mainly focused on collect and summarizing of documentation such as transcripts, articles, papers, and books from more than 25 source, regarding the importance of public open spaces for the community. The summarizing content analysis of this paper is the fundament for a qualitative oriented procedure of text interpretation used to analyse the information gathered. Results of this study confirms that sound social interaction need an appropriate physical space including criteria of: design, activities, access and linkage, administration and maintenance, place attachment and users' characteristics, also previous studies in this area have a health perspective with measures of physical activity of open spaces in general.

Keywords: social interaction, open spaces, natural environments, outdoor recreation, content analysis method.

1. Introduction

Social interaction also can be described as an interchangeable sequence of dynamic exchanges through which individuals can attach meaning, interpret, and respond. People who report having more satisfying and active social lives tend to report feeling better about themselves and their lives and peoples how enjoyable interactions were self-assured, life satisfaction, and psychological well-being [21]. While open spaces one of the most important areas in the city, it provides different recreation and sport activities for children and aged, it also enhances physical, social, mental and environmental components for cities and citizens [5].

Content from this work may be used under the terms of the Creative Commons Attribution 3.0 licence. Any further distribution of this work must maintain attribution to the author(s) and the title of the work, journal citation and DOI.
Published under licence by IOP Publishing Ltd

Urban public open space is also a main component of urban fabric that caters to the urban public daily life. The amount of green space in a neighbourhood is significantly correlated with the health of its residents [6]. Social interaction is a common social issues that exist in in most societies nowadays, so it has to be highlighted this issue and try to deal with it in urban and architectural perspective. According to previous literatures and studies to provide a brief on significance of open space in social interaction and recreation. Almost most studies confirm that social interaction need an appropriate place to occur, also there are different factors affecting social interaction in the green open space such as the design, components, distance, location and facilities of open space [9]. Types and nature of recreational and physical activities within this space also enhance the role of open space in social interaction. So the objective of this paper is to identify factors affecting social interaction of users in green open spaces in order to provide proper foundations and standards for open space design through implementing qualitative method to answer the following questions:

I. What are the criteria of green open spaces to enhance social interaction?

II. How can achieve social interaction in the urban fabric?

2. Social Interaction and Green Open Space

The development of social relationships in urban scale should begin with careful attention to specific spaces by implementing and using interaction design approaches, specific spaces also used to effectively address the sociocultural relationships of the people [19]. So to achieve the social interaction must provide physical space, thus parks have crucial roles to develop and enhancement cities as well as social relations, so it's not only a place where people join the nature but also communicate each other socially and culturally [15]. Open spaces wherever people are present, other people are attracted to that location and get together, move among each other and looking for a place for themselves among others, then new activities begin. Looking, listening and talking are the activities which determine the quality of open space. Good space with good facilities plays an important role in the formation of such activities and elements. These spaces reinforce social contacts and meet their psychological needs of residents [23].

Elements of social open space include: walking trails, horse trails, gardens, children's play areas, sitting areas, and green space. While open space at all levels in the hierarchy plays a role in providing open space opportunities for the surrounding community, while the hierarchy of open space levels ranging from: the neighborhood space, through the sub district (a catchment of three neighborhoods), district (usually containing sports facilities, a catchment of six neighborhoods) and regionally spaces to the state level spaces [7-11]. Pocket park is a small outdoor space, provide a safe and inviting environment for surrounding community members, it also meet a variety of needs and functions e.g. small event space, play areas for children, spaces for relaxing or meeting friends, taking lunch breaks, etc. [7]. Social relationships also vary according to the age, gender and place, where place element is a key factor to determine the nature of social activity for all ages. While public spaces allow people to meet on ostensibly neutral ground in planned and unplanned ways, to interact with others within the context of the whole community, so public spaces can contribute to the cohesion of communities [12]. Individual usage and social interactions in public spaces are influenced by many factors, including how the spaces connect and the design, maintenance and management of the natural and the built environment.

Bekker describe that how to create a social communication in open space through interactive play objects by describing some design values that apply for designing playful interactions for social interaction and physical play [2]. His study highlights that community (players) in recreational zone (recreational parks) can create a wide range of physical games and being shared and participated as a type of social interaction (fig. 1). The importance of this study also concentrate on its aims which invite people especially children to replace the television, computers and video games with physical games that participate with other people in open space to stimulate social interaction [2]. The woodcarving, components and decoration also have considered as an important component in traditional Malay vernacular architecture and it was playing their appointed roles in the complex network of social relations and the value system of the Malays so decoration and motifs of the spaces influence special social interaction [26].

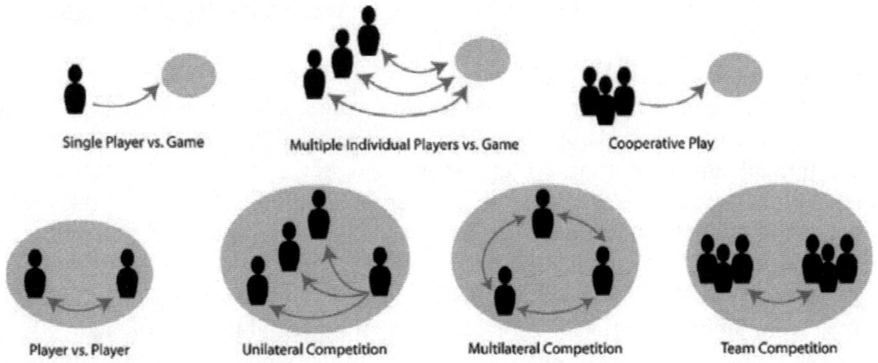

Figure 1. Patterns of Players Interaction.
Source: Bekker et al., 2010

3. Criteria for Enhance Green Open Spaces

Layout plan and design of outdoor spaces has a great role to lead individuals to outside spaces, socializing and being together, also space design tools can be used to provide social interaction and the feeling of being a community especially in transformation places [30]. "Community" is one of the most important concepts taken up in connection with surroundings of the house, so "a good society designers" believe in the necessity of changing the spatial structure in order to change life and it is possible to create a new social order in the city through spatial arrangements [30]. While complex social diseases can be cured through different house surroundings, parks and new spatial arrangements [30]. Open spaces are also a kind of crystallization of social life where the citizens are living. The inherent behaviour is closely associated with the environment, so organizing human behaviour and its application in time and space, is the most important factor in designing open spaces. Human being is the main component of this space which credits the space by their values and norms [23]. The socio-demographic characteristics of a neighbourhood affect how residents interact with each other in outdoor spaces. Factors such as: age, marital status, and presence of children at home, owner-renter status, length of residence, educational attainment and annual income are relevant socio-demographic characteristics presumably associated with social interaction [10].
While to achieve successful public park, good of access and linkage (GAL) should be the main factor in designing a park follow by sociability (SOC), user and activities (UAC) and degree of comfort and image (DCI) factors [26]. Park is an important space in the relationship of man and nature to promote and provide space for physical activity, health behaviour, and can reduce some diseases such as diabetes and certain cancer, also parks provides the benefits of community interrelationship, as well as increase the value of the property, so public parks have always been an important component in an urban area [26]. A high-quality outdoor space can enhance social interaction by attracting people to come and stay for some time. The more time people spend in outdoor space, the more likely are they to engage in activities [16]. The existence of interesting objects, features or elements, such as artificial water scenery and properly arranged seats, children playgrounds and recreational facilities also encourages the use of public space and enhances social bonding among residents [13-16].
With the shift to a service based society, providing opportunities for outdoor recreation that enables mental and physiological self-regulation has become an increasingly important landscape function [3]. The quality of the nearby recreation area is at least an as important condition for these benefits as the easy access to these areas [3]. Inhabitants' satisfaction with the recreation area, their activity and interaction level within the recreation areas as well as the time spent in the recreation areas appeared to

be more relevant predictors for these benefits than the frequency of visits in these areas. While recent research has provided considerable evidence that visits to near natural everyday landscapes promote psychological, well-being and physical health [3]. Holland suggests that security and safety are an important factor in open spaces especially for older people, while for others risk-taking is essential to ways of living [12].

The role of place attachment in making places more meaningful, also the emotions and a sense of pride keep the place as the center for personal and social engagement for social interaction [29]. While the role of place attachment is important in park utilization and social interaction (were conducted to examine the role of place attachment in park utilization and social interaction, so the study found that frequent visits, proximity to residences, and the landscape features contribute to the development of attachment to a place) [29]. The significance of the park as a social integrator could be unfulfilled if the functional attachment to a place is weak. Urban spaces and places should belong to the public, therefore, they should be shaped for the good of the people physically and emotionally. The green spaces function as community parks in the city have a great potential to develop a social environment. The parks should be inclusive and meaningful to the life of the members of the community amidst the egocentric nature of the modern society. Green recreational places are the most important places to generate happy and healthy urban inhabitants [29].

4. Benefits of Green Open Space

Godbey considers that how being outside in natural surroundings may improve health and how outdoor physical activities benefit participants [9]. Particular attention is given to children's health problems that can be mitigated through outdoor play, sports, and nature study [9]. His study also confirmed a link between physical activity that takes place outdoors and positive health outcomes and also an association between an indoor, sedentary lifestyle and negative health consequences. There is also evidence that both being outdoors and viewing natural scenes can reduce stress [9]. Outdoor recreational activity is helping to mitigate health consequences associated with sedentary behavior. Enhanced understanding of socio-demographic differences in physical activity (PA) location preferences could therefore contribute to health promotion [18]. While the environment surrounding residences and its recreational and commuting opportunities are believed to affect human health and well-being, provided scientific evidence for the mechanisms of influence of the types of environments on human well-being, this study examined how the presence of and access to green spaces is related to the level of physical activity and self-rated health (fig. 2) [22]. The presence of an access to green space is evident in the suburbs, where outdoor recreation was related to leisure time physical activity and to self-rated health. While easy accessibility to green spaces should be an important objective in the management and planning of urban and suburban forests and other green spaces [22].

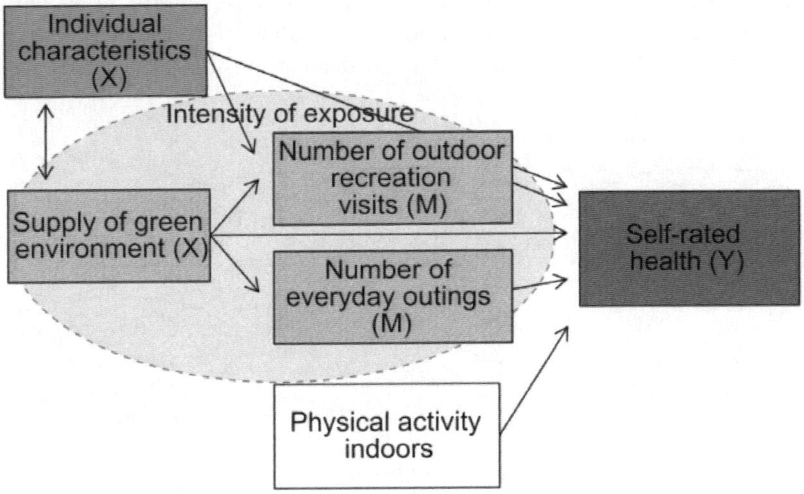

Figure 2. Environments and Human Well-being

Some studies also indicate that the decline of social life in housing estates is closely related to the design of communal outdoor spaces, while the spatial arrangement of residential area has been found to enhance social interaction among residents and influence their activity patterns [1]. Walking around a small group of houses to sit in a confined space were significantly related to social interaction and friendship formation. While close-to-home recreation opportunities are particularly vital for older people [1]. There is a need for better collaboration between park and protected area and health institutions, and effectively integrate empirical evidence into relevant policy, planning and management, where there are a positive links between human contact with nature and health and well-being. The role of parks also can be summarized in term of five different components: physical, mental, spiritual, social and environmental [24]. Open spaces are also providing opportunities for growth, creativity, fosters social interaction such as: communication, relaxation, entertainment and a place for circulating and cohesion residential site, are quite effective in reducing pollution and improving the environment [23].

5. Factors Affecting Green Open Spaces

Climate change is one of the most pressing global environmental issues facing the world today and one that has major social, economic and environmental repercussions [27]. The majority of visitors are concerned about climate change in a national park they visit, and would support agency efforts to mitigate possible climate change effects [27]. So it should contemplate differences in perceptions of the effects of climate change and travel behavior according to visitor characteristics e.g. age, gender and market segments. Other studies state that urban parks built in the early period of the cities were designed as places fully integrated into the urban fabric, where people from all walks of life can use freely, encounter others, learn from each other and develop, the parks built under the current ideology aren't [8]. With the changing of the existing ideology, the urban parks which were built to design a modern urban life have become highly commercialized and vehicles to obtain economic benefits. Owing to their social, economic, political, and ideological meanings, public spaces are convenient tools for urban renewal strategies [8]. While according to Buta attachment must be discussed as having implications for pro-environmental behaviors directed at protecting natural resources and informing sustainable recreation and ecosystem management [4]. Place attachment plays a determining role in residents' behavior toward engaging with the park. They also mention the importance of understanding the social environment within communities proximal to a park and how social perceptions can shape attachment to a protected area with implications for pro-environmental civic engagement beliefs and intentions to sustain natural resource protection, where there is a strong relationship between community attachment, behavioral beliefs and behavioral intentions [4].

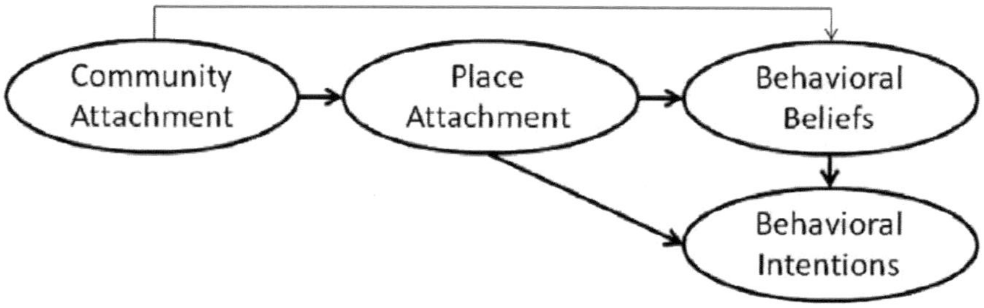

Figure 3. Mediation of Place Attachment

6. Methodology of the Paper

The main objective of the paper is to identify the factors that affecting social interaction in green open spaces, through proving open spaces with an appropriate criteria are important to enhance social interaction and community. So it is necessary to answer the following research questions: 1) what are the criteria of green open spaces to enhance social interaction? and 2) how can achieve social interaction in the urban fabric?

This paper employs (Mayring, 2000) recommendation to design the method of the paper while data obtained from more than 25 source varied among articles, researches and books, regarding the importance of open spaces and green area for the community and interaction. Summarizing content analysis is a tool of research methodology used to determine the presence and meaning of concepts, terms, or words in one or more form or structure of the information. This systematic and replicable technique allows for compressing many words of text into fewer content categories on explicit rules of coding in order to allow researchers to make inferences about the author (Mayring, 2000).

7. Discussion

Results of the content analysis regarding to the previous studies about the importance of green open spaces to enhance social interaction through implementing an appropriate criteria of this spaces. So findings of this study attempt to answer the paper questions and achieve its objectives. Finding of the study confirms that public spaces allow people to meet on ostensibly neutral ground in planned and unplanned ways, to interact with each other within the context of the whole community, then public spaces can contribute to the cohesion of communities, also the criteria of that open space could determine the nature of social activity and interaction. Previous literatures, researches and studies have different opinion about the key factors and criteria of open space to enhance communities; for some good facilities and elements plays an important role in the formation of open space, others state that design and layout have a great role to lead individuals to outside spaces [30].

Users' characteristics in open spaces are also an important criteria including; age, marital status, presence of children at home, educational attainment and annual income. While good of access and linkage should be the main factor in open spaces design follow by sociability, activities and degree of comfort are the key factors according to [26]. A high-quality outdoor space also can enhance social interaction while nearby recreation area with easy access is at least an as important factor of open spaces [3] also role of place attachment is important to make open places more meaningful. While security and safety are an important factor in open spaces [12]. So individual usage and social interactions in open spaces are influenced by varied and different criteria, the following criteria identify an answer for the first question of the paper about the criteria of green open spaces to enhance social interaction:

i. Design, scenery and image of the open space.
ii. Elements and quality of open space.
iii. Activity, and play object of the open space.
iv. Good access, connect and linkage of the open space.
v. Users' characteristics and interest.
vi. Place attachment of the open space.
vii. Administration and maintenance of the open place.
viii. Safety and security of the open space.

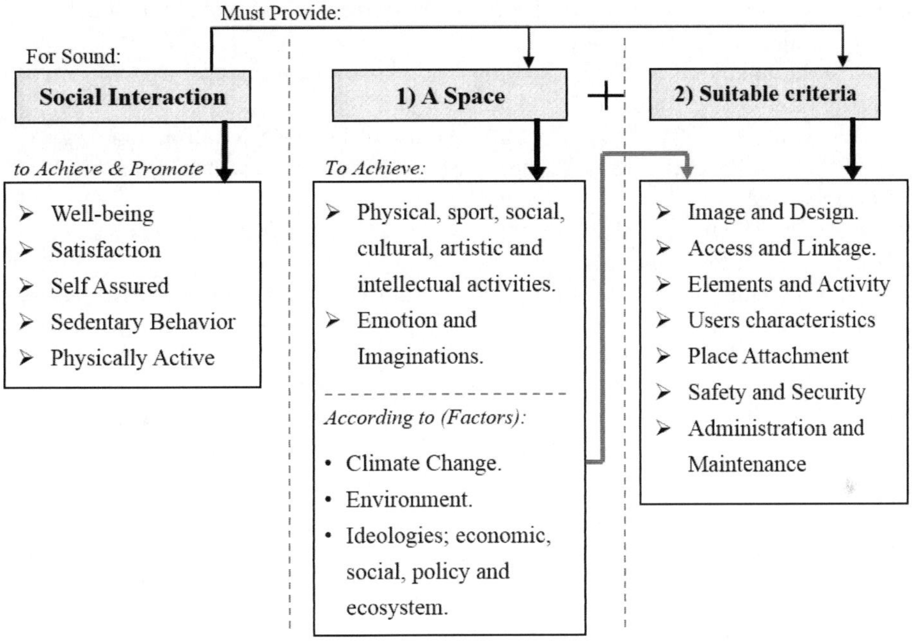

Figure 4. Means to enhance social interaction in Urban Fabric
Source: S Salih & S Ismail, 2017

8. Conclusion

The research results indicate that one of the means to achieve social interaction is to provide an appropriate space. Criteria and components of green open space are also an important mean to achieve sound social interaction including, design, activities, elements, access, users' characteristics, security, administration and maintenance and place attachment of open spaces and parks. It has been mentioned the benefits and importance of social interaction in green open spaces as well, including; improve health and wellbeing, sedentary behavior and reduce stress, enhance socio-demographic differences, urban renewal strategies, economic improvement and raise the value of property.
Therefore the results of this study prove that achieving green open space is an essential to enhance physical, health, mental, social, environmental, and aesthetic aspects for residents and residential area, thereby encourages social interactions in the city between citizens through implementing an appropriate criteria and components in that space. The results of this paper are useful reference for urban and landscape planners, architects, social psychologist, and researchers in this fields. That will be in the line with the aim of this paper to contribute in outdoor recreation and relaxation development fields and highlight the role of open spaces in social development, so the significance of this paper lies in linking social-psychology to architectural research.

9. References

[1] Abu-Ghazzeh T. 1999. Housing layout, social interaction, and the place of contact in Abu-Nuseir, Jordan. *Journal of environmental psychology*. **19**(1): 41-73.

[2] Bekker T., Sturm J. and Eggen B. 2010. Designing playful interactions for social interaction and physical play. *Journal of Personal and Ubiquitous Computing*. **14**: 385-396.

[3] Buchecker M. and Degenhardt B. 2015. The effects of urban inhabitants' nearby outdoor recreation on their well-being and their psychological resilience. *Journal of Outdoor Recreation and Tourism*. **10**: 55–62.

[4] Buta N., Holland S. and Kaplanidou K. 2014. Local communities and protected areas: The mediating role of place attachment for pro-environmental civic engagement. *Tourism Recreation Research*. **5**(6): 1-10.

[5] Carlisle R. and Stankovich A. M. 2014. Role of Parks, Recreation, and Open Space in Age-Friendly Place making. *MRPA Professional*. pp. 6-3.

[6] Chen Y., Liu T. and Liu W. 2016. Increasing the use of large-scale public open spaces: A case study of the North Central Axis Square in Shenzhen, China. *Journal of Habitat International*. **53**: 66-77.

[7] Cooper M., C. and Francis C. 1998. *People Places: Design guidelines for urban open space*. 2nd Ed. John Wiley & Sons Inc. Canada.

[8] Ekinic Z. and Sağlam H. 2016. Meanings and social roles of the republic period urban parks in Ankara. *Journal of Procedia Social and Behavioural Sciences*. **216**: 610-621.

[9] Godbey G. 2009. *Outdoor Recreation, Health, and Wellness: Understanding and Enhancing the Relationship*. Resources for the Future. Washington, USA.

[10] Haggerty, L. 1982. Differential social contact in urban neighborhoods: environmental vs. sociodemographic explanations. *The sociological quarterly*. **23**(3): 359-372.

[11] HCC. 2015. *Hume City Council: Open Space Strategy*. Hume City Council. Hume. pp. 1-89.

[12] Holland C., Clark A., Katz and Peace S. 2007. *Social interactions in urban public places*. Joseph Rowntree Foundation. Bristol.

[13] Huang S. 2006. A study of outdoor interactional spaces in high-rise housing. *Landscape and Urban Planning*. **78**(3): 193–204.

[14] Isaac S. and Michael W. B. 1997. *Handbook in research and evaluation: A collection of principles, methods, and strategies useful in the planning, design, and evaluation of studies in education and the behavioral sciences*. 3rd Ed. Education and Industrial Testing Services. San Diego.

[15] Kara B., Tuncay H. and Deziz B. 2011. Investigating recreational qualities of the parks in Aydın. *Procedia Social and Behavioral Sciences*. **19**: 158-164.

[16] Knack E. 2000. Hanging out: teens search for the perfect public space. *Planning Journal*. **66**(8): 4–9.

[17] Krippendorf K. 1980. *Content analysis: An introduction to its methodology*. Sage Publisher. Beverly Hills, CA.

[18] Larson L., Whiting J., Green G. and Bowker J. 2014. Physical activity locations in Georgia: Frequency of use by socio-demographic group. *Journal of Outdoor Recreation and Tourism*. **5**(6): 68-72.

[19] Mamaghani M., Asadollahi A., and Mortezaei S. 2015. Designing for Improving Social Relationship with Interaction Design Approach. *Procedia - Social and Behavioral Sciences*. **201**: 377-385.

[20] Mayring P. 2014. *Qualitative content analysis: theoretical foundation, basic procedures and software solution*. Misc. Klagenfurt.

[21] Nezlek J., Richardson D., Green L. and Jones E. 2002. Psychological well-being and day-to-day social interaction among older adults. *Journal of Personal Relationships*. **2**: 1350-4126.

[22] Pietial M., Neuvonen, Borodulin, Korpela K., Sievanen and Tyrvanen L. 2015. Relationships between exposure to urban green spaces, physical activity and self-rated health. *Journal of Outdoor Recreation and Tourism.* **10**: 44–54.

[23] Poodeh S. and Vali A. 2014. Investigating the Characteristics of Open Spaces to Enhance Social Interactions in Neighbourhood Environments. *Journal of Natural and Social Sciences.* **3**(4): 148-158.

[24] Romagosa, Eagles P. and Lemieux. 2015. from the inside out to the outside in: Exploring the role of parks and protected areas as providers of human health and well-being. *Journal of Outdoor Recreation and Tourism.* **10**: 70–77.

[25] Sagalyn L. and Johnson A. 2014. Entertainment Centres and the Quest for Transformation in the Periphery. *disP - The Planning Review.* **50**: 18-37.

[26] Sakip S., Akhir and Omer S. 2014. Determinant Factors of Successful Public Parks in Malaysia. *Procedia Social and Behavioral Sciences.* **170**: 422-432.

[27] Stone S., Scaccia M. and Poteet D. 2015. Exploring visitor perceptions of the influence of climate change on tourism at Acadia National Park, Maine. *Journal of Outdoor Recreation and Tourism.* **11**: 34-43.

[28] Tan T. 2008. Entertainment is Emotion: The Functional Architecture of the Entertainment Experience. *Media Psychology.* **10**: 55-62.

[29] Ujang, Moulay A. and Zakariya K. 2015. Sense of Well-Being Indicators: Attachment to public parks in Putrajaya, Malaysia. *Procedia - Social and Behavioral Sciences.* **202**: 487-494.

[30] Uslu A. and Gökçe. 2010. Social interaction in urban transformation areas and the characteristics of urban outdoor spaces. *African Journal of Agriculture Research.* **5**(20): 2801-2810.

IOP Conf. Series: Materials Science and Engineering **291** (2017) 012002 doi:10.1088/1757-899X/291/1/012002

The Barriers and Causes of Building Information Modelling Usage for Interior Design Industry

A B Abd Hamid[1], M Z Mohd Taib[2], A H N Abdul Razak[3] and M R Embi[4]

[1&2]Centre of Interior Architecture Studies, Faculty of Architecture, Planning and Surveying, Universiti Teknologi MARA, Puncak Alam Campus, Selangor Malaysia,
[3]Department of Interior Design, Faculty of Architecture, Planning and Surveying, Universiti Teknologi MARA, Seri Iskandar Campus, Perak, Malaysia
[4]Department of Architecture, Faculty of Built Environment, Universiti Teknologi Malaysia, Johor Malaysia

E-mail: abah7591@salam.uitm.edu.my

Abstract. Building Information Modeling (BIM) has since developed alongside the improvement in the construction industry, purposely to simulate the design, management, construction and documentation. It facilitates and monitors the construction through visualization and emphasizes on various inputs to virtually design and construct a building using specific software. This study aims to identify and elaborate barriers of BIM usage in interior design industry in Malaysia. This study is initiated with a pilot survey utilising sixteen respondents that has been randomly chosen. Respondents are attached with interior design firms that are registered by Lembaga Arkitek Malaysia (LAM). The research findings are expected to provide significant information to encourage BIM adoption among interior design firms.

1. Introduction

At present, the use of technology in the construction industry continues to thrive and develop. Various fields including architecture, engineering, construction, quantity survey, facility management including interior design are shifting to benefit the vast development in technology. Building Information Modelling (BIM) is an example of the technology that is being adopted in these fields and interior design, among others are moving towards utilizing it as an effective tool to fulfil the interior design industry. A recognizable design field across the globe, interior designers are specialized in interior space's detailed and specification design obligations closely related to the selection on material and finishes, lighting setting, colour, texture, ventilation, furniture, ergonomic, anthropometric and all related with the human factor and purpose [1, 2]. Therefore, interior design industry are moving forward and ready to implement Building Information Modelling (BIM) as a tool or process to balance the demands of industry at present.

Interior design is a recognize able design field which is specialized in interior space's detailed and specification design obligations which is related to the selection on material and finishes, lighting setting, colour, texture, ventilation, furniture, ergonomic, anthropometric and all related with the human factor and purpose [1, 2]. These details put the special characteristic for interior design with others profession. This study will focus on the importance of using BIM in the interior design industry

Content from this work may be used under the terms of the Creative Commons Attribution 3.0 licence. Any further distribution of this work must maintain attribution to the author(s) and the title of the work, journal citation and DOI.
Published under licence by IOP Publishing Ltd

will be evaluated the effectiveness and implementation of the design and documentation process. Besides, the direction of interior design industry will be determined and the barriers that inhibit the using of BIM in the design industry cannot be implemented. Interior design is still struggles to using the technology interior design is loose a grip in design field instead of having the upper hand [3].

This study will focus on the significance of BIM in the interior design industry that will be evaluated based on its implementation in the design and documentation process. Besides, future direction of interior design industry and the barriers that inhibit the use of BIM shall be determined. Interior designers are lagging in the technology race and gradually slipping behind in the design field where, they could have the upper hand advantage [3]. BIM can be described as a tool and process that functions as a building information modelling and management that characterizes represent; document and manages information; inbuilt intelligence, analyze and simulates, and collaborates and integrates [4]. Most of industry fields such as architecture, engineering, construction (AEC); facility management, quantity surveying and etc have already implemented BIM as main tool and process to assist and expedite tasks and various activities.

2. Literature Review

2.1. Interior Design and BIM

Interior design is multi-faceted profession in which creative and technical solution is applied within its own domain to achieve a better built environment. These solutions enhance the quality of life and culture of the occupants, and are aesthetically attractive. According to [5] depicted that, interior designers are required to have a good understanding on construction detailing to ensure effective use BIM. Since the modelling is holistic, interior designers must understand the core and shell of building details in order to design effectively in BIM.

Interior design profession describes a building's internal space with accurate scale to manipulate design elements such as wall, ceiling and floor. Interior designer's concern is the human ergonomic that ensures working and functional space. According to [6], interior design is a professional body that focuses on creative and technical to solve design problems.

2.2. A Theory and Practice of BIM

Building Information Modelling (BIM) is a revolution in design practice and standards of building design, delivery and operation. The purpose of BIM doesn't only focus on Computer Aided Design (CAD), but it is an innovative and landmark technology that is spectacularly attractive to allow for a better performance in design [7]. BIM improves the management of human skills and organisation by diminishing delays and inaccuracy between a project's stakeholders. In practice, it is an improvement in the design and monitoring of projects by taking into consideration the lowered cost, improved accuracy, speed and safety elements.

BIM provide opportunities for different and multiple fields to collaborate and engage data exchange in relevant areas, values, and commercial aspect in construction disciplines [8, 9]. Furthermore, this technology is able to grant the competitive advantage by creating opportunities and lowering the costs [10] than is showing as function of Technology, Process and People [11]. An interior design process is initiated with several design options, elaborated and developed driven by varieties of factor; space utilization, area requirements, aesthetics, material cost, finishes and material – which is applied in decision-making process. With the use of such technology, the documentation of each phase will ease up the process that will enable faster progress of the project

Table 1. Understanding the Anatomy of BIM

<table>
<tr><th>No.</th><th>Technology</th><th>No.</th><th>Process</th></tr>
<tr><td>1.</td><td>Project simulation consists using of 3D models integrated with project planning, design, construction and operation [12].</td><td rowspan="2">1.</td><td rowspan="2">Refining and adjusting according to project specifications and design changes to ensure model accuracy [14].</td></tr>
<tr><td>2.</td><td>Introduced the object-oriented parametric modeling [13].</td></tr>
<tr><td>3.</td><td>The process where the element is modified or assembly automatically [15].</td><td rowspan="2">2.</td><td rowspan="2">To produce a design with optimized for quality, aesthetics, constructability, affordability, timeliness and seamless flow into lifecycle management [18].</td></tr>
<tr><td>4.</td><td>Defined BIM as a supplier, operation and maintenance producers, and clearance [16, 17].</td></tr>
</table>

Table 2. BIM for Interior Design

<table>
<tr><th colspan="3">BIM for Interior Design</th></tr>
<tr><td>Interior Advantages</td><td colspan="2">1. Speed and ease of interior design creation and its ability to visualize the design
2. Ability to capture and manage the design via multiple options – from space layout to material selections.
3. The richness and reliability of data - early fundamental task.</td></tr>
<tr><td>Interior Design Options</td><td colspan="2">1. To keep multiple design alternative until enough information decision making.
2. Enabling designers to develop and study multiple design alternatives simultaneously within a single model.</td></tr>
<tr><td rowspan="3">Interior Design Information</td><td>a. Schematic design</td><td>- Able to create a master schedule for interior spaces.
- A working database view of a project.
- Able to input table-based schedules for many interior design components.
- Produce accurate preliminary estimates.</td></tr>
<tr><td>b. Calculating Quantities</td><td>- Storage, manage and coordinates design
- Calculate area and quantify material and reliably – reducing overall project cost
- Accurate area measurement – carpet, wall coverings, ceiling tile, or finishes.
- Ability to track occupancy, space allocation and assets.</td></tr>
<tr><td>c. Interior Documentation</td><td>- All the construction documents relating to the interior design are direct presentation.</td></tr>
</table>

3. Research Aim and Methodology

This study aims to identify and elaborate barriers of BIM usage in interior design industry in Malaysia. The preliminary study is to obtain the initial data concerning barriers adopting BIM in interior design industry. This study focuses on ID firms registered with Lembaga Arkitek Malaysia (LAM) which sixteen ID firms in Klang Valley were randomly chosen. Beginning with find the research topic based on the literature review and preliminary study following with the questionnaire, discussion, analysis and finding to find out the problems and issues.

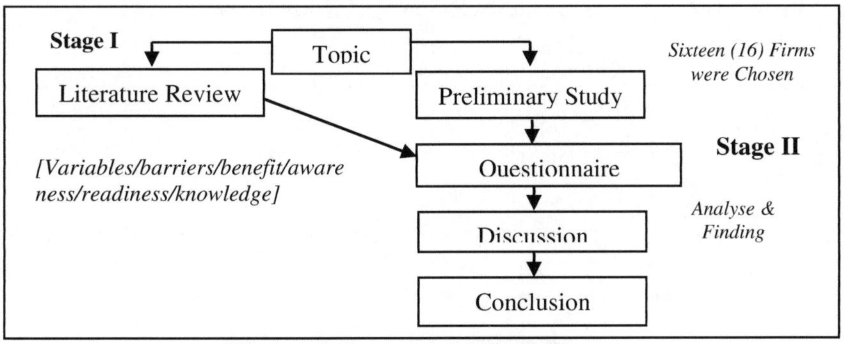

Figure 1. Research Method Flow

4. Result Analysis

Based on the analysis, this research will highlight the barriers and causes of interior design industry still behind in BIM adoption. From the analysis, this study will identify the awareness, readiness, problems and barriers of BIM appropriate usage in the interior design firms.

Table 3 shows the responses of year of firm established for interior design. Accordingly in Table 3, 6 respondents representing 38% of the total firms established between the year 2006 and 2010 while 5 respondents representing 31% of the total firms established between 2011 and 2015. It is followed by 3 respondents representing 19% of firms established between 2001and 2005. However, between the year 1990 and 1995 and 1996 and 2000 each represented 1 respondent (6%) only.

Table 3. Year of Firm Established

Year	Frequency	Percentage
1990-1995	1	6%
1996-2000	1	6%
2001-2005	3	19%
2006-2010	6	38%
2011-2015	5	31%
Total	16	100%

Table 4 is shows the awareness level of interior design using BIM. It revealed 7 respondents representing 44% having poor knowledge about BIM while 5 respondents representing 31% have fair knowledge about BIM. However, only 4 respondents representing 25% have very poor awareness about BIM awareness

Table 4. Level of Awareness and Readiness BIM for Interior Design

Level	(VP) 1	(P) 2	(F) 3	(G) 4	(VG) 5	Total
Awareness Level	4	7	5	0	0	16
Readiness Level	4	0	10	2	0	16

Note: VP=Very Poor, P=Poor, F=Fair, G=Good, VG=Very Good

Table 5 shows the level of knowledge instead of experience of BIM. From the table 5, four (4) respondents representing 25% of the total respondents know about BIM while twelve (12) respondents representing 75% of the total respondents do not know or never heard of BIM.

Table 5. Level of Knowledge

Response	Frequency	Percentage
Yes	4	25%
No	12	75%
Total	16	100%

Figure 2 shows the ranking of barriers to the application of BIM in interior design industry in Malaysia. The most critical part is technology with 18 responses with each 17 responses in IT and knowledge elements as a barriers for using BIM. The others with 16 responses each are awareness, readiness, time and financial.

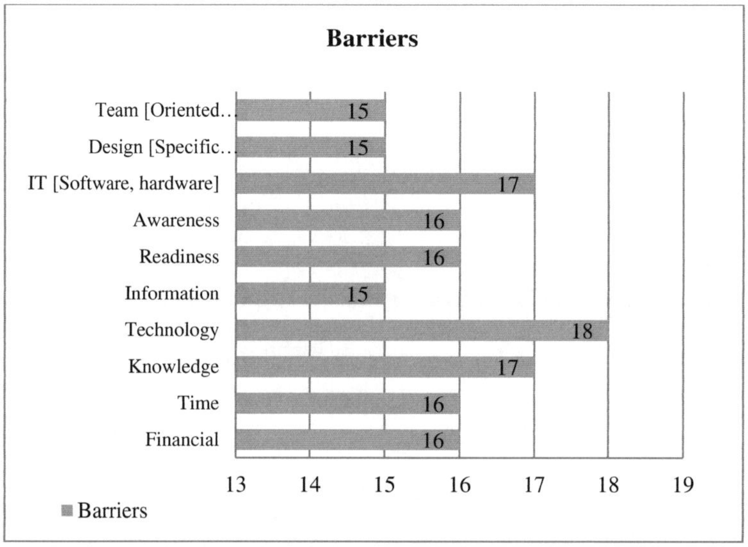

Figure 2. Ranking of the Barriers to the application or use of BIM in ID

Figure 3 shows the reasoning a few firms using BIM. Only 2 respondents using BIM where is 1 respondent representing 6% for each respondent and 14 respondents representing 88% did not apply the BIM.

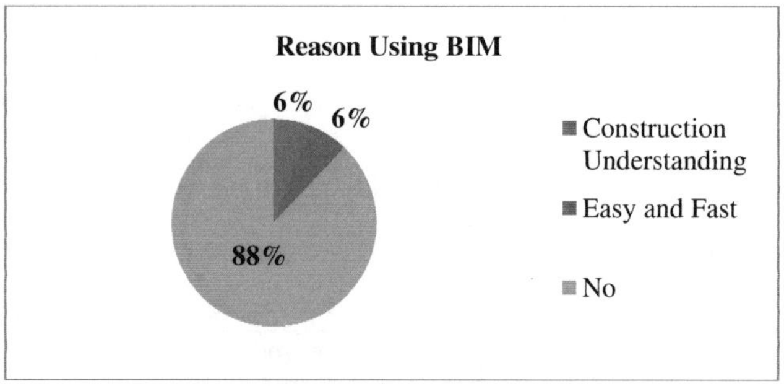

Figure 3. Reason Using BIM

Figure 4 shows the respondents the reasoning not using BIM in interior design firm. According to figure 4, 30% is representing less exposure about BIM and 21% are no information about BIM and paid to consultant. Otherwise, 21% respondents are said they are preferred using software 3D and 7% said no useful for company and lack of staff.

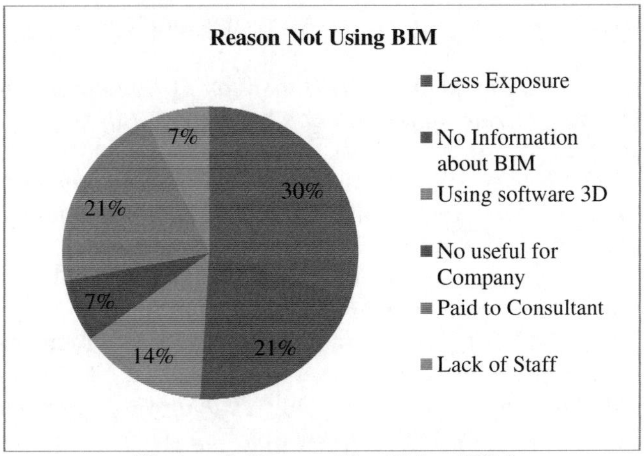

Figure 4. Reason Not Using BIM

5. Discussion and Conclusion

BIM is very important in the construction industry to enhance the quality of work, time, cost and documentation. Therefore, interior design industry must look forward to using BIM even there are barriers to facing. The crucial challenges to adopt BIM faced by the interior design industry in Malaysia context have been focused in this study and several influent factors have been identified. The most critical are caused due to the factors which are determined by technology, knowledge, cost, awareness and readiness.

BIM application in interior design projects can reduce cost and time in the design progress to sustain the quality of projects. BIM benefits for interior design industry by improving design phases and progression, decreasing the cost and maintenance, and also improving the skill and knowledge between human resource and employee. Implementation BIM in the interior design industry can enhance the overall quality of skill and improve the interior design profession as part of construction players. Moreover, with BIM technology will be expected to implement widely in the interior design industry in Malaysia

Acknowledgments

Authors would like to acknowledge the Universiti Teknologi MARA (UiTM) and Ministry of High Education (MOHE) for the financial support under the Lestari Grant. The authors also would like to thanks to whom assist along the research progress for contributing part of this research.

References

[1] Kurtich J Eakin G 1993 *Interior Architecture* John Wiley & Sons Inc. New York.
[2] Coles J House N 2007 *The Fundamentals of Interior Architecture* Ava Publishing House West Sussex.
[3] Burcu S Burcu G 2012 *A Domain Specific Models for Interior Architectural Education and Practice* Automation in Construction 21 Elsevier 10-23.
[4] Ning G Vishal S Kerry L 2014 *BIM Ecosystem: The Convolution of Products, Processes and People* 15 Building Information Modeling: BIM in Current and Future Practice.
[5] Crumpton A Miller B 2008 *Building Information Modeling: State of the A&D Industry and BIM into Design Education.* In Proceedings of the Interior Design Educator's Council Annual Conference 1175–1188.

[6] National Council for Interior Design Qualification 2000 *NCIDQ Examination Study Guide.* Washington, DC: NCIDQ, pp. 22–23.
[7] Karen M K 2014 *Introduction: On the Theory and Practice of BIM, Building Information Modeling: BIM in Current and Future Practice*, John Wiley and Sons Inc Hoboken New Jersey

[8] Tse T K Wong A K and Wong F K 2005 *The Utilisation of Building Information Models In nD Modelling: A Study of Data Interfacing and Adoption Barriers, Electronic Journal of Information Technology in Construction*, **vol. 10**, pp. 85-110.
[9] Gao J and Martin F 2008 *Framework & Case Studies Comparing Implementations & Impacts of 3D/4D Modeling Across Projects.* Technical Report Number 172 Centre for Integrated Facility Engineering, University of Salford, UK.
[10] Mitsutaka H 1989 *Evaluation of Technology in Construction.* Technical Report Number 16 Centre for Integrated Facility Engineering, University of Salford, UK.
[11] Koskela L 1992 *Application for the New Production Philosophy* Technical Report Number 72, Centre for Integrated Facility Engineering, University of Salford, UK.
[12]Kymmell W 2008 *Building Information Modeling: Planning and Managing Projects with 4D CAD and Simulations USA* McGraw Hill Construction.
[13] Azhar S Nadeem, A Mok, J Y N and Leung B H Y 2008 *Building Information Modeling (BIM): A New Paradigm for Visual Interactive Modelling and Simulation for Construction Projects'* Proceedings of the First International Conference on Construction in Developing Countries (ICCIDC-I), Karachi, Pakistan, August 4-5.
[14] Jorge C and Kathleen I October 2007 *BIM: Who, What, How and Why Building Operating Management*
[15] Daniel J S 2011 *Design Integration Using Autodesk Revit® 2012 Mission* KS: SDC Publications.
[16] Azhar S and Richter S 2009 *Building Information Modelling (BIM): Case Studies and Return-on-Investment Analysis* Proceedings of the Fifth International Conference on Construction in the 21st Century (CITC-V), Istanbul, Turkey, 1378-1386.
[17] CRC Construction Innovation 2007 *Adopting BIM for Facilities Management: Solutions for Managing the Sydney Opera House* Cooperative Research Centre for Construction Innovation, Brisbane, Australia.
[18] McGraw-Hill Construction 2008 *Building Information Modelling: Transforming Design and Construction to Achieve Greater Industry Productivity*, McGraw-Hill Construction, New York,

Communication, coordination and cooperation in construction projects: business environment and human behaviours

Wesam Salah Alaloul[1], Mohd Shahir Liew[2], and Noor Amila Wan Zawawi[3]

[1, 2 &3] Department of Civil and Environmental Engineering, Universiti Teknologi PETRONAS, 32610 Seri Iskandar, Perak, Malaysia

wesamalool@hotmail.com

Abstract. The accomplishment of construction projects is extremely dependent on the integration of several stakeholders; therefore none of them has the control or the ability to accomplish the project alone. Each of them may influence and be influenced by the project management approach. There is no comprehensive theoretical platform for defining Communication, Coordination and Cooperation (3Cs) in the management of construction project. This paper deliberates the function of the 3Cs different theoretical perceptions. Through an analysis of selected articles from reputable academic journals in construction management, the business environment and human behaviour were identified as two main parts. A little has been done so far about the 3Cs, and how they are correlated with construction projects performance. Therefore, the objective of this paper is to explain the definitions and the association between the 3Cs. There is a significant link between communication and coordination. Coordination alternatively, is trust-based a logic of mutual and exchange. Consequently, cooperation is much more sophisticated, which needing more time and attempts.

1. Introduction

The construction industry suffers from fragmentation during the project lifecycle, as activities are subdivided into further detailed levels. This context has been considered as a serious cause for poor performance, low efficiency, and bad competitiveness. Construction projects have several stakeholders, such as client, designer, engineers, contractor, and subcontractors whom should deal together on the project procurement and execution [1]. Management of construction projects includes the integration of three critical flows of materials, information and cash between the stakeholders. The modern business context described by complex and competitive which forced construction stakeholders to create effective and efficient management schemes. During the last decade, there has been a fast tendency in the direction of main initiatives to enhance the construction projects performance. Therefore using new terms and particularly as an operational context, Communication, Coordination and Cooperation (3Cs) were developed. The construction industry nature creates difficulty in deciding what alternative of 3Cs approaches could be valuable in a given conditions or tasks [2]. Due to this gap in the literature, this paper aims to investigate the existing literature related to 3Cs concepts. The 3Cs can be used interchangeably and are assumed the understanding of what stakeholders should do to work together efficiently. The 3Cs are organised procedures to offer a harmony of actions in the completion of shared objective, while specifies combined efforts of actors working in an organization. It is expected that when organizations grow, so make the 3Cs approaches they applied. However, those approaches decrease

Content from this work may be used under the terms of the Creative Commons Attribution 3.0 licence. Any further distribution of this work must maintain attribution to the author(s) and the title of the work, journal citation and DOI.
Published under licence by IOP Publishing Ltd

ambiguity by generating the chance of a smaller period of accurate feedback and by aggregate information validity [3]. The 3Cs could even decrease the amount of information processing, by exchange the applicable information only on time and to the appropriate stakeholder. Finally, Communication without cooperation has no fruit and cooperation without coordination has no root.

2. Literature Review

There have been no specific definitions of the terms Communication, Coordination and Cooperation (3Cs) even though they appeared in academic and practice frequently. However, the descriptions of the 3Cs differ amongst the sections of literature based on the paper source and discipline [4]. To the best of the available information, emphasis on the outlines of information exchange and long-term activities integration. A limited current research, conversely, has started to explore how communication, coordination and cooperation subjects correlate, and the way they individually or together affect project performance. The 3Cs enables a systematic management to project planning, organizing, scheduling, and control to explore the relations between stakeholders a construction project. Table 1 shows the definitions for the three terms given by famous dictionaries.

Table 1. Definitions for the 3Cs terms

Dictionary	Merriam-Webster [5]	Cambridge [6]
Communication	"A process by which information is exchanged between individuals through a common system of symbols, signs, or behaviour."	"Two-way process of reaching mutual understanding, in which participants not only exchange information and ideas but also create and share meaning."
Coordination	"The harmonious functioning of parts for effective results."	"The act of making actors involved in plan activity work together in an organized way."
Cooperation	"To work jointly with others or together especially in an intellectual endeavour."	"To work jointly with others or together especially in an intellectual endeavour."

It could be concluded that several definitions are applied to describe 3Cs terms; they did not provide fixed and exact features of one term against the other. The definitions indict that the 3Cs are a sequence and growth of relationships started by communication to reach the cooperation with continuous coordination. Communication is the approach of linking actors together. In management, it is a vital role; therefore, an organization could not work without communication among its hierarchy levels and units. It refers to how actors contact each other and the way information (not just "facts," but policies, prospects, feelings, and ideas) is transmitted. However, "lack of communication" is at the top of problems list in most organizations. Most of the actors are better at sending information out than receiving it in. The instrument of communication and response represent an essential portion of coordination and cooperation, so communication features have a significant effect on the project performance [7]. The shared information should be accurate as possible with shortest and simplest way. However, communication is challenging for many causes. Terminology and processes applied to exchange information differ between different organizations. There is an indecision to rely on other organizations, because of the lack of understanding or personal disagreements. Also, there is no consensus on the gathering and distributing information or to whom it must be transferred [8].
Coordination is the combination of various portions into an arranged system to accomplish the predefined objectives. It is an unseen power that connects all other functions of management. Based on Xue, et al. [1], "Coordination is the act of managing interdependencies between activities to achieve a goal". In an inter-organizational perspective, coordination can be defined as the thoughtful and organised arrangement of stakeholders' activities to accomplish the predetermined objectives. Coordination is typically accomplished with efficient communication or information transfer. It is

assumed as the connecting, meshing, harmonisation, or arrangement of tasks. Coordination appearances to notify every unit in the project as to how and when it should be performed. Coordination normally includes the requirement and information distribution, decisions, and feedback instruments to unify and carry instruction needed for resources optimisation. Coordination pursues to guarantee that stakeholders' efforts yield the optimum results with fewer procedure losses. Therefore, it considers the level to which every unit in the project endeavour considers other units considerations. The absence of coordination could lead to an incapability control, confusions between stakeholders, inefficiency in resources usage, and replication of work [1]. The motivation for applying coordination is the reality of dependencies among tasks so that the aim is to handle dependencies to make the tasks as portions of determined objectives. Management normally mentions to the necessity of project separation into work packages. However, separation of project rises effectiveness, it generates or at least rises, the necessity for coordination. Coordination can be considered as effective once the stakeholders are satisfied. Malone and Crowston [1994] note that "good coordination is nearly invisible, and coordination is notice clearly when it is lacking". All project stakeholders face coordination challenges since by definition they involve some division of project and some task interdependence between them. Amongst the main coordination difficulties in any organization is the one between the main office and field units and cultural differences between stakeholders [9].

Cooperation is now a hallmark not only of every management process but also as a culture. Cooperation is the procedure of shared the achievement: two or more actors with harmonising abilities contacting to establish a shared objectives which none had earlier controlled. Cooperation from a mutual sense regarding a procedure, product, or a task. Different from communication, to transfer information, it is about applying information to generate something none of the cooperators can achieve alone. Moreover, the opposite of cooperative is competitive. Different from coordination, cooperation aims for different understanding and spontaneity, not only structural harmony. As the cooperation relationship represent the last stage of the 3Cs chain, its operational needs the formation of an earlier communication and continues coordinative relations [10].

Cooperation mentions to long-term prescribed relations, like outsourcing or commissioning. For instance, in a real cooperative association, cooperation agreements would be established after some level of sharing predictions and exchanges concerning. Nevertheless, as associations reach the cooperative phase, these agreements grow to include further strategic features like long-term contracts and knowledge investments [11].

3. Methodology

This study provides a systematic review of reputable journals to track the growth of the 3Cs in the construction industry. The investigation applied for this study started with a meta-analysis of project management exploration and literature on 3Cs crossing all disciplines. Concentration was located in the present literature in construction management articles, precisely those with an emphasis on business context and human behaviours. Known the fast growing environment of this arena, the investigation highlighted articles finalized in the previous ten years. The article's sources have been recognised as the protuberant and significant-quality sources and have been engaged in reviewing other subjects in construction. For instance, Lin and Shen 2007 [12], using a keywords exploration in similar sources, offered a critical review of evaluating the performance of value management research in the construction industry. Also, Chan et al.2002, [13] identified the critical factors of design/build and partnering construction projects. In this study, the articles were investigated by keywords, title and abstract. Several terms have been applied to define 3Cs, like "collaborative, alliance, partnership, partnering, networking, coalition, consortium, teamwork, joint venture, community, and coordination". Consequently, the way to choice the correct keywords to retrieve the articles is an essential concern that could have a large influence on the investigation outcomes. To complete the investigation; first, a preliminary exploration was conducted by all of the previous terms. The outcomes illustrations that more than 300 paper were retrieved. In order to enhance the efficiency of the review procedure, the titles and keywords of the 300 paper were checked and concluded that "communication, coordination, and cooperation" are the most relevant keywords. There is 180 paper that involves at least one of the keywords. Finally, the processes for retrieving the articles were as below: (1) the titles, keywords, and abstracts were checked with the keywords in every article. (2) Abstracts were evaluated to realize whether these articles fitted for the study purposes. (3) After sieving, 34 paper reporting on 3Cs were used and scanned in details for the purpose of this study.

4. Results and Dissection

The fluctuating construction industry business is considered as extreme competitiveness, and complex context necessitates construction organizations to start effective management schemes [1]. The 3Cs have been regarded as critical factors for organizations existence and performance enhancement in the construction industry. The intention of separating among cooperation, coordination, and communication and well-establish organizations is problematic, especially in construction management. They accomplish what the stakeholders aim to effect composed as integrated potential. The 3Cs have the benefits that attract the stakeholders, for instance, improve the possibility of enhanced performance or objectives achievement. This attraction is due to, 3Cs not only share the effort depend on every actor knowledge but accomplish a continuous information exchange. Enhancing efficiency of organizational procedure and the necessity for further well-organized usage of the resources are two issues that motivate the 3Cs.

The significance of communication is due to its capability to allow stakeholders to work composed on shared activates or toward common objectives. The organization that engaged communications only between stakeholders, where everybody is socialized to norms, does not certainly have high performing projects. The absent portion is the linking among the stakeholders' efforts to coordinate their efforts. The analysis of cooperation and coordination associated investigation so far has presented that most of them studied cooperation and coordination subjects separately. A group of research suggest that the degree of cooperation within an association could affect its components and accordingly the worth of the coordination and vice versa [14]. On the other hand, the probability that cooperation and coordination connected association features outline performance interactively rather than independently. Coordination could also endorse further cooperation that is wide and extensive. Cooperation and coordination effect each other and have a mutual influence on project results that surpasses their joint effects. The connections of the 3Cs are displayed in figure 1:

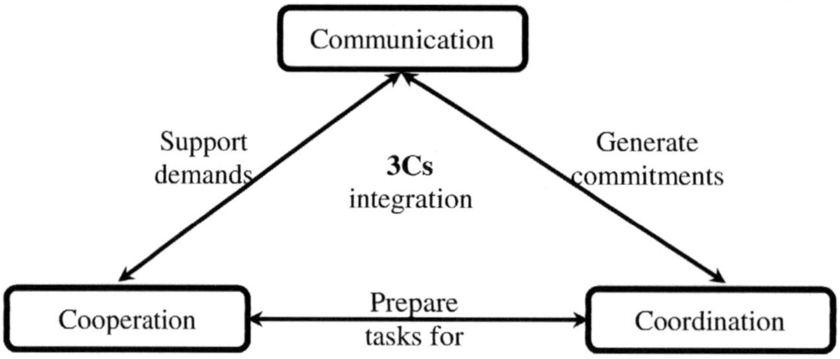

Figure 1. The 3Cs triangle relating communication, coordination and cooperation

Figure 1 shows that the communication needs stakeholders' commitment to become coordination and coordination prepare the tasks for cooperation. However, the level of communication should be increased to support the demands of cooperation. Therefore, the 3Cs can be interpreted individually as; Communication is considered as the exchange of information. It hinges on the selection of media, transmission approaches, discussion arrangement and tracks. Coordination is indicated by the harmonicity of stakeholders, their tasks and resources. It is the controlling perception of integration.

Cooperation is the fruit that can be gained in a shared working context. It is the workflow or project lifecycle perception. The arrangement can be applied on storyboard interface [15].
A conceptual focus on coordination issues rather than cooperation issues provides a different perspective on relationships: while the cooperation perspective centres attention on partners' level of agreement, the contribution of resources, and the sharing of benefits [16]. The coordination perspective highlights the specific ways that stakeholders devise to implement and manage the relationship. Importantly, coordination challenges are not automatically resolved because stakeholders' interests align. On the contrary, coordination problems require intelligent, vigorous, persistent and organized effort. Table 2 presents a comparison between the 3Cs based on the "Vision and Relationships, Structure and Responsibilities, Authority, Accountability and Resources Management".
Human behaviours have direct influences on 3Cs success in construction projects. Behaviour analysis has become an important research area to investigate projects performance improvement.

Table 2. A comparison between the 3Cs characteristics.

Element	Commination	Coordination	Cooperation
Vision and Relationships	Organizational mission and objectives are not deliberated. Interaction is as needed basis/no time limit.	Mission and goals are deliberated for compatibility. Interface typically on particular project.	Shared, novel mission and objectives are formed. One or more projects are started for longer-term outcomes.
Structure and Responsibilities	Relations are informal; depend on every organization purposes individually. Information is exchanged as required.	Organizations assume required functions but still individually. Channels for contact are formed.	Novel structure and/or formal division are formed. Several stages of channels are formed.
Authority and Accountability	Authority rests uniquely with individually. All authority and responsibility rests with every organization.	Authority rests on every organization, but there is coordination. There is some shared risk, but most still individually.	Authority is defended by the stakeholders to equilibrium ownership. The equal risk is shared.
Resources	Resources are separate.	Resources are available for a specific project.	Resources are pooled for a long-term effort.

Trust has been founded as the most significant influence on effective 3Cs, which is not only the main factor of behaviours but also a critical factor for a business strategy.

4.1. OVERVIEW OF 3CS IN CONSTRUCTION PROJECTS

In Fact, 3Cs in construction projects have emerged and implemented for a long period from the viewpoint of working together. The development of 3Cs can be seen from two viewpoints: a working relationship and delivery systems, as shown in figure 2.

IOP Conf. Series: Materials Science and Engineering **291** (2017) 012003 doi:10.1088/1757-899X/291/1/012003

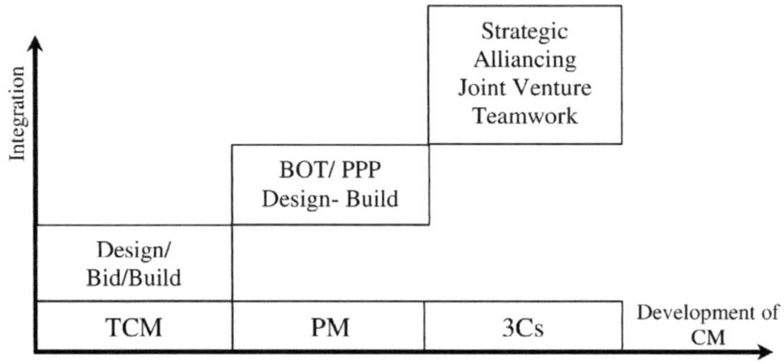

Figure 2. Development of 3Cs in construction projects

From the viewpoint of delivery systems, the construction industry mainly experienced three development processes: the TCM strategy, the PM strategy, and the partnering strategy. TCM relies on independent firms brought together by competitive bids and tough contracts. From the viewpoint of working relationship, there are various forms of 3Cs in construction projects, such as partnership, alliance, joint venture, and coalition. The central principle of these forms of 3Cs is to build an integrated team or entity virtual or physical to work integrally with the predefined objectives. These 3Cs forms have been applied to improve performance and working relationships in construction projects.

5. CONCLUSION

The business context of construction projects is characterized by adversarial relationships, fragmented operation and complexity. The 3Cs business culture, attitudes, and working strategy in the construction industry are emerging to improve the project's performance. Research on the business context of 3Cs in construction projects mainly focuses on the design and construction stage at the project or organizational level. Although the volume of alliance activity is growing, the potential benefits of such arrangements are not being realized in practice yet. This is because the stakeholders are emphasizing short-term profit sharing and accountancy procedures to the detriment of the more long-term value-added dimensions. The purpose of this paper is to address the current gap in project management literature, which need the adequate definition of 3Cs. More work is required for establishing the boundaries of 3Cs theoretical approaches, about construction projects performance.

References

[1] X. Xue, X. Li, Q. Shen, and Y. Wang, "An agent-based framework for supply chain coordination in construction," Automation in construction, vol. 14, pp. 413-430, 2005.

[2] M. B. W. Fritz, S. Narasimhan, and H.-S. Rhee, "Communication and coordination in the virtual office," Journal of Management Information Systems, vol. 14, pp. 7-28, 1998.

[3] J. H. Miller and S. Moser, "Communication and coordination," Complexity, vol. 9, pp. 31-40, 2004.

[4] T. Ellingsen and R. Östling, "When does communication improve coordination?," The American Economic Review, vol. 100, pp. 1695-1724, 2010.

[5] Merriam-Webster, Merriam-Webster's collegiate dictionary: Merriam-Webster, 2004.

[6] D. Matsumoto, The Cambridge dictionary of psychology: Cambridge University Press, 2009.

[7] R. Barkhi, A. Amiri, and T. L. James, "A study of communication and coordination in collaborative software development," Journal of Global Information Technology Management, vol. 9, pp. 44-61, 2006.

[8] H. Fuks, A. Raposo, M. A. Gerosa, M. Pimentel, D. Filippo, and C. Lucena, "Inter-and intra-relationships between communication coordination and cooperation in the scope of the 3C

IOP Conf. Series: Materials Science and Engineering **291** (2017) 012003 doi:10.1088/1757-899X/291/1/012003

Collaboration Model," in Computer Supported Cooperative Work in Design, 2008. CSCWD 2008. 12th International Conference on, 2008, pp. 148-153.
[9] W. Alaloul, M. Liew, and N. Zawawi, "Coordination process in construction projects management," in Engineering Challenges for Sustainable Future: Proceedings of the 3rd International Conference on Civil, Offshore and Environmental Engineering (ICCOEE 2016, Malaysia, 15-17 Aug 2016), 2016, p. 149.
[10] M. Cataldo, P. A. Wagstrom, J. D. Herbsleb, and K. M. Carley, "Identification of coordination requirements: implications for the Design of collaboration and awareness tools," in Proceedings of the 2006 20th anniversary conference on Computer supported cooperative work, 2006, pp. 353-362.
[11] J. J. Mohr, R. J. Fisher, and J. R. Nevin, "Collaborative communication in interfirm relationships: moderating effects of integration and control," the Journal of Marketing, pp. 103-115, 1996.
[12] G. Lin and Q. Shen, "Measuring the performance of value management studies in construction: critical review," Journal of Management in Engineering, vol. 23, pp. 2-9, 2007.
[13] A. P. Chan, D. Scott, and E. W. Lam, "Framework of success criteria for design/build projects," Journal of management in engineering, vol. 18, pp. 120-128, 2002.
[14] W. S. Alaloul, M. S. Liew, and N. A. B. Zawawi, "A Framework for Coordination Process into Construction Projects," in MATEC Web of Conferences, 2016, p. 00079.
[15] W. S. Alaloul, M. S. Liew, and N. A. W. A. Zawawi, "Identification of coordination factors affecting building projects performance," Alexandria Engineering Journal, vol. 55, pp. 2689-2698, 2016.
[16] W. S. Alaloul, M. S. Liew, and N. A. W. A. Zawawi, "The characteristics of coordination process in construction projects," in Technology Management and Emerging Technologies (ISTMET), 2015 International Symposium on, 2015, pp. 159-164.

IOP Conf. Series: Materials Science and Engineering **291** (2017) 012004 doi:10.1088/1757-899X/291/1/012004

Towards Implementation of Green Technology in Sabah Construction Industry

Noor Azland Jainudin [1, a)], Ivy Jugah [2, b)], Awang Nasrizal Awang Ali [3, c)] and Rudy Tawie [4, d)]

[1,3,4] Faculty of Civil Engineering, Universiti Teknologi MARA, 94300, Kota Samarahan, Sarawak.

[2] Faculty of Administrative Science and Policies Studies, Universiti Teknologi MARA, 94300, Kota Samarahan, Sarawak.

[a)] azland9824@sarawak.uitm.edu.my
[b)] ivyjugah@sarawak.uitm.edu.my
[c)] awang295@sarawak.uitm.edu.my
[d)] rudy@sarawak.uitm.edu.my

Abstract. The construction industry in Sabah is one of the major roles for development of social, economic infrastructures and buildings in generating wealth to the state besides the tourism sector. The increasing number of construction projects particularly in the rapid developing city of Kota Kinabalu, green technology as a whole is becoming more significant as it helps to develop effective solutions to encounter global environmental issues. The objective of the research is to identify the awareness and implementation of green technology in construction industry in Kota Kinabalu, Sabah. The methodology of the research is through distributing the questionnaire to the contractors, developers, consultants, architects and state government agencies to the area in Kota Kinabalu only. The questionnaires had been analysed to find out the mean value. 100 questionnaires distributed to the respondents but merely 85 questionnaires collected have been analysed. Based on the findings, 83.5% organisations were aware with the concept of green technology in construction project. In terms of the implementation only 64.7% had been implemented in their organizations. More than 50% from the major players such as contractors, consultants, developers, architects and state government agencies were aware based on six green technology concepts in their organizations. As a conclusion, the awareness towards green policy concept in construction industry is very satisfied. Meanwhile, in terms of implementation need to be increased the number of organizations to be involved in green technology in construction industry.

Keywords: Awareness, implementation, green technology and construction industry.

1. Introduction

Construction industry is a major consumer of non-renewable resources and a massive producer of waste, and the operation of buildings is responsible for around half of the total CO2 emissions [1]. The

Content from this work may be used under the terms of the Creative Commons Attribution 3.0 licence. Any further distribution of this work must maintain attribution to the author(s) and the title of the work, journal citation and DOI.
Published under licence by IOP Publishing Ltd

technology can be in the form of green buildings or renewable energy such as hydro, solar and biogas. The implementation of sustainable methods in this industry is required to achieve win-win outcomes as to attain mutual benefit for the environment, the advanced society, and the economy of the industry [2]. Although there has been lengthy discussion on the green technology, the visibility of the implementation is still yet to be seen widely in Kota Kinabalu. On 29th January 2008, Sabah Development Corridor (SDC) was launched to enhance and accelerate the growth of Sabah's economy for a better quality of life, promote regional balance and bridging the rural-urban divide while ensuring sustainable management of the states' resources [3]. The implementation of green technology is more significant as it helps to develop effective solutions to encounter global environmental issues because currently increasing number of construction projects especially in the rapid developing city of Kota Kinabalu.

In Malaysia, the urgency to build green is mandated by our Prime Minister to reduce the global greenhouse gas emission by 30% in the year 2020. This is also mainly driven by factors such as to fulfill the market demands, to provide comfort and health for the occupant, save more money and also to act as an environmental responsibility. There are already a few discussions on the viability for its implementation in Malaysia [4, 5]. There are many development projects in Sabah and Sarawak particularly in the rapid developing Kota Kinabalu areas. What makes it interesting is to know whether construction players such as contractors, developers, consultants, architects and state government agencies are all aware of the green technology concept because it is believed that if all the construction practitioners are being well-groomed with the importance of implementing green concept, then their projects will also be integrated with green technology system. The construction projects have an effective impact on economy, environment, human health and the conventional system cannot meet the demands of a multifunction building [6]. Currently, only several studies have been conducted to investigate the implementation of green concept in Kota Kinabalu construction industry.

Green technology is an initiative to improve various kinds of systems and materials, from techniques for generating energy to non-toxic products [4]. Application of green technology is a major requirement to achieve the developed country status, in term of design and construct sustainable projects that can provide energy, water savings and healthier indoor environment in their projects. As a responsibility to the future generations, it is important to understand the modern concept when designing the buildings and to reduce the negative impacts of environment. The key criteria when implementing green technology include Energy Efficiency (EE), Indoor Environment Quality (EQ), Sustainable Site Planning and Management (SM), Materials and Resources (MR), Water Efficiency (WE) and Innovation (IN) [7]. The developers need to be encouraged to consider the environmental quality of the end-users of their construction products. In order to improve cost-saving productivity, natural resources need to be used effectively [8].

2. Research Methodology

The study is focus on the awareness and implementation of green technology in construction industry. Therefore, it must be conducted with the limited amount of resources but the study can be scoped only in Kota Kinabalu area. Two types of data, that is primary data and secondary data. For the primary data, the questionnaires have been conducted and distributed to the various key players such as developers, contractors, state government agencies, architectural and others. For the secondary data, literature reviews are written from references of articles, journals and from the internet that are related to the study.

100 questionnaires distributed and sent to the respondents but only 85 questionnaires successfully collected due to the several limitations from the respondents. Then, the collected data analyzed by using SPSS program and determine the percentage of each element of the awareness and implementation of green technology in construction industry.

This questionnaire method is practical and the ideas from the respondent can be easily assessed. The questionnaire consists of several sections. The first section is actually to determine the percentage of awareness and implementation of green technology in construction industry. Next section is problem identification, using Likert Scales. The final section of the questionnaire is the strategies for implementation which to be used to find out ways to speed up implementation of green technology.

3. Result and Discussion

3.1. Awareness and Implementation of Green Technology

Figure 1 shows that the awareness of the organization on the green policy concept. The finding shows that the familiarity of the organization towards the concept of Green Technology in construction project is 83.5%. Only 16.5% of the respondents stated that their organization were not familiar with the concept. Figure 2 shows that the implementation of green technology in construction industry in Kota Kinabalu. The data shows that, majority of the respondents, 64.7% stated that their organization had been implemented the concept of Green Technology in construction project and only 35.3% stated that their organization has not implemented the concept of Green Technology.

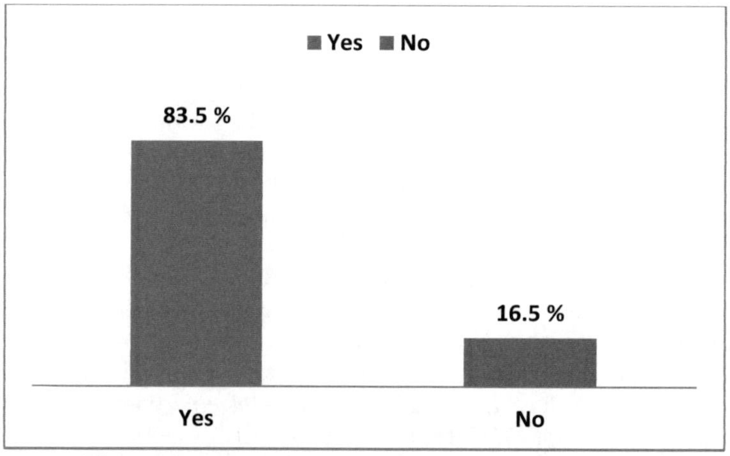

Figure 1: Awareness towards green policy concept

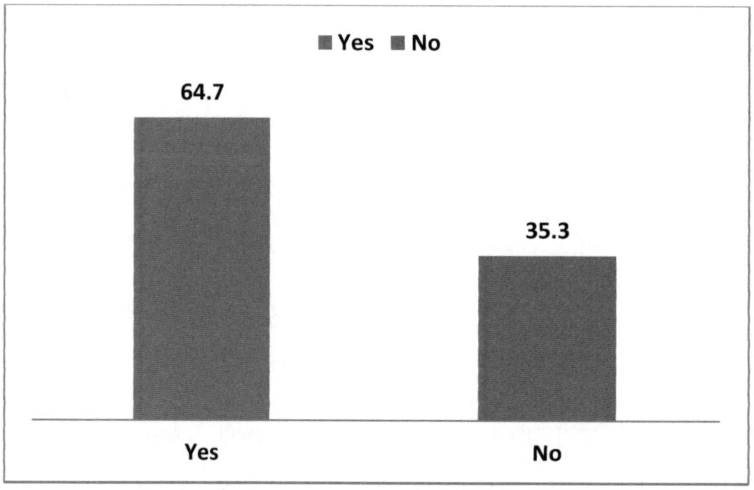

Figure 2: The implementation of green technology

3.2. *Six Concepts of Green Technology in Organization: Awareness and Implementation*

There are six concepts of green technology discussed in this study as shows in Figure 3. The first concept of green technology is energy efficiency whereby it is refer to a process of reducing energy consumption by using solar heat through the optimization of building orientation, solar energy harvesting and ensuring regular building maintenance. The finding shows that 56.5% of the respondents stated that their organization were aware on the concept of green technology in term of energy efficiency. Next, only 41.2% of the respondents stated that their organization were aware and has implemented the concept of green technology. Merely 2.4% of the respondents stated that their organization were not aware and not implement the concept.

Secondly, indoor environmental quality concept in an organization refer to a good indoor quality performance in terms of visual and comfort, use quality air filtration, and proper control of temperature and humidity in an organization. This study shows 58.8% of the respondents stated that their organization aware on this concept and 35.3% of respondents' organization aware and implement. Thirdly, sustainable site planning and management is another important concept in green technology. Meanwhile, awareness on conserving environmental area through replantation, implementing proper construction method and storm water management is a must in any organization. The finding shows, organization managed to implement sustainable site planning and management in a workplace. About 47.1% of the organization aware and implement this concept. Only 3.5% of the organization are not aware and not implement the concept of selecting sustainable site planning and management. Any organization must also promote the use of environmental-friendly materials from sustainable and recyclable source. The data shows majority of the organizations (57.6 %) are aware and about 40% has practice on the use of environmental-friendly materials in an organization.

In addition, an organization should have the fittings for rainwater harvesting to recycle water for various usages because this initiative can create awareness on green technology in a workplace. The finding shows that 40% of the organizations are aware and implement on this practice in a workplace. However, only 1.2% indicates that their organizations were not aware and not implement the water efficiency on the concept of providing water efficiency in a workplace. Lastly, innovative design and initiatives that use Green Technology, renewable energy and environmental-friendly product such as green parking lot. This study shows that 69.4% of the organizations aware on innovative concept but only 8.2% of the organization have implemented the innovative concept of green technology in their organizations. However, 22.4% respondents said that their organizations were not aware and not implement the innovative on green technology. This is the highest percentage among of the six green technology concepts which is the organizations were not aware and not implement.

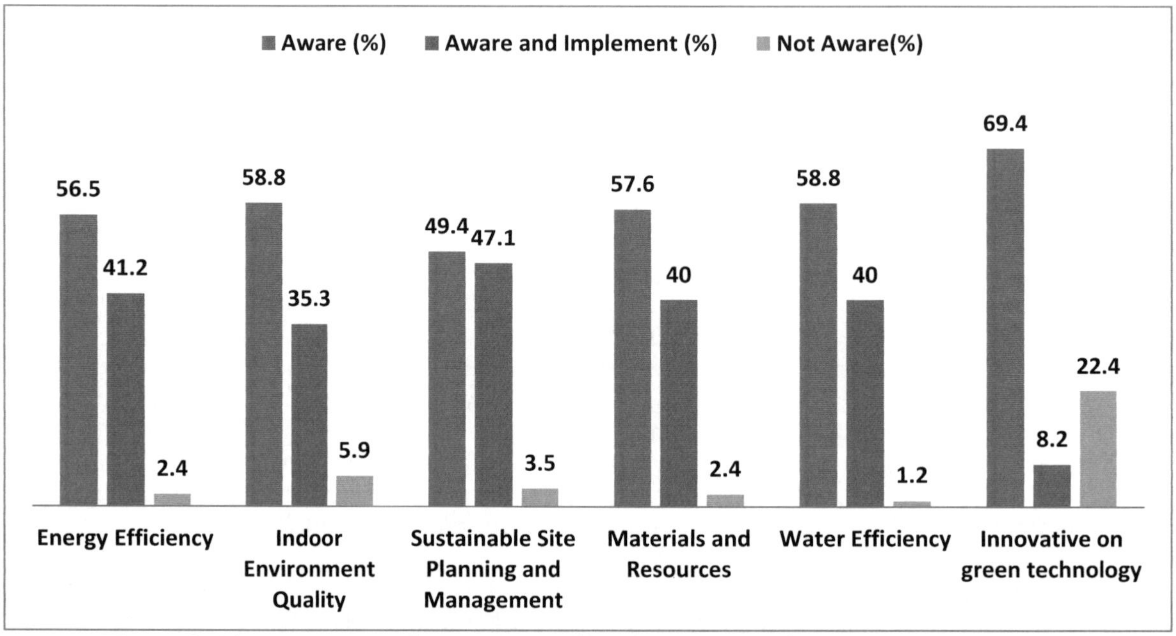

Figure 3: The percentage of awareness and implementation of green technology concept

4. Conclusion

Based on the data analyses, 57.6% of the respondents stated that their organization were aware on the concept of green technology in term of materials and resources. Only 2.4% of the respondents stated that their organization were not aware and not implement the concept for both energy efficiency and materials & resources. Based on the finding, more than 50% for six green technology concepts from the major players such as contractors, consultants, developers, architects and state government agencies were aware of the green technology concept in their organizations. Meanwhile, in terms of the organizations is not aware and not implement the green technology concepts are less than 5% except innovative on green technology 22.4%. That means, majority of the organizations in Kota Kinabalu aware and implemented on green technology concepts.

5. Acknowledgments

The authors are thankful to Geran Dana Pembudayaan Penyelidikan (RAGS) for funding this paper, Ministry of Higher Education (MOHE), Research Management Institute (RMI) Universiti Teknologi MARA (UiTM) for all the valuable supports. Also, special thanks to Universiti Teknologi MARA Sarawak that gave a lot of information, encouragement and support during this research.

IOP Conf. Series: Materials Science and Engineering **291** (2017) 012004 doi:10.1088/1757-899X/291/1/012004

References

[1] Zuhairi, A. H., Kamarul, A. M. K., Mohd, K. G., Maria, Z. M. and Ahmad, Z. A. R. (2011). The Construction Industry Perspective and Current Innnitiative, 10th Annual Conference and Meeting, IIUM, Kuala Lumpur.

[2] Yoong H. Chan, Brenda C. T. Lee, Jin C. Lee (2014). World Academy of Science, Engineering and Technology International Journal of Social, Behavioral, Educational, Economic, Business and Industrial Engineering Vol:8, No:4.

[3] SEDIA, (2011). Sabah Development Corridor. Retrieved December 16, 2012, from http://www.sedia.com.my.

[4] Bakar, K. A., Sam, M. F. M., Tahir, M. N. H., Rajiani, I., and Muslan, N. (2011). Green technology compliance in Malaysia for sustainable business development. Journal of Global Management, 2, 1, pp. 55-65.

[5] Huat, N. B., and Akasah, Z. A. (2011). An overview of Malaysia green technology corporation office building: A showcase energy-efficient building project in Malaysia. Journal of Sustainable Development, 4, 5, pp. 212-228.

[6] Deng, S., Dai, Y. J., Wang, R. Z., and Zhai, X. Q. (2011). Case study of green energy system design for a multifunction building in campus. Sustainable Cities and Society, 1, pp. 152-163.

[7] Sood, S. M., Chua, K. H., and Peng, L. Y. (2011). Sustainable development in the building sector: Green building framework in Malaysia. In 15th International Conference on ISO & TQM (15-ICIT), Malaysia.

[8] Parkin, S. (2000). Sustainable development: The concept and the practical challenge. Proceedings of ICE, 138, pp. 3-8.

IOP Conf. Series: Materials Science and Engineering **291** (2017) 012005 doi:10.1088/1757-899X/291/1/012005

Colour Light And Wellbeing: A Case Study Of M Mall 020 George Town, Penang Island

Safial Aqbar Zakaria[1], Ng Yii Rou[1], Hoi Zhi En[1], Tai Iyian[1]
[1]Interior Design Department, Universiti Sains Malaysia, Minden, Pulau Pinang, Malaysia

ssafial@usm.my, joying@gmail.com, zhi_en0801@hotmail.com, linluoxuan1997@gmail.com

Abstract. Contrary to popular belief, the brightest light or the most colourful light makes good lighting design. However, what makes a good lighting design in interior space is an impeccable composite of art and science. With the application of good lighting design, it can produce an impressive result from enhancing the aesthetic of architectural elements to conveying the right type of ambiance of the interiors. This research intends to address the crucial issues regarding the ways in which lighting designers can communicate the benefits of good lighting and to create a better awareness to users. The objectives of this paper are to outline and explore the features of good and poor lighting design in M Mall O2O based on the lighting design language and profession. The results of this research are mainly qualitative in nature, supported by the professional lighting designers on the definitions of good lighting, personal observation and visual data which were taken in George Town, Penang Island. The case studies on good and poor lighting portrayed in this mall were used as examples to scrutinize the issues raised herein. To achieve the optimum lighting design, a joint approach of focusing on the artistic flair brought forth by lighting and more scientific effort on the calculation levels of lights is crucial. Different functionality requires a different amount of attention on either approach. In conclusion, a good lighting design must be able to enhance the atmosphere and also enrich the quality of the interior architecture. Apart from that, a good lighting design should have good distribution of brightness levels, contrast and different colour temperatures to enhance characters of the interior spaces without neglecting the health and wellbeing aspects.

1. Introduction

Lighting is one of the important aspects of an interior space as it provides light, cast and brightness to the users in order to carry out the daily activities. In case, it has the ability to change the mood of a space just as it does the perceived size of a space. Lighting is defined as the use of various light source which has a particular quality, to achieve some aesthetic or practical effect while illuminating a scene. Light can be obtained through the natural daylighting as well as the artificial lighting. The source of all daylights is the sun, that brings in or captures the daylight in an interior space with an opening requirement such as windows and skylight. In particular, the aim of having a daylighting design in a building is to maximize visual comfort or reduce energy usage. Thus, artificial lighting is introduced to tackle this problem. The artificial lighting refers to the light source supplied by man-made equipment which emits light and heat. Both natural daylighting and artificial lighting are fundamental and act as the key element for architecture and interior design. Lighting fixtures, usually associated with interior design, light fixture or luminaire, are electrical devices that use electric lamp to create artificial light. Contemporarily, lighting is utilized for both aesthetic and function purposes. For aesthetic purpose, it is used as decorative lighting especially for an interior design and also on a building "skin". The main reason for using this artificial light is it can become an eye–catching interior and also create a soothing environment. Simultaneously, in terms of functionality aspect, lighting is required by the users to perform the tasks. Nevertheless, with the rise in awareness of health aroused among the global society, people have started to link lighting with health. Also, they have started doing research and investigated how lighting can be influential to one's health. Thus, all types of research have been done by inventors, they also cooperate with scientists to acquire sufficient knowledge on lighting for the scientific aspect. Eventually, the findings display various health issues

Content from this work may be used under the terms of the Creative Commons Attribution 3.0 licence. Any further distribution of this work must maintain attribution to the author(s) and the title of the work, journal citation and DOI.
Published under licence by IOP Publishing Ltd

caused by poor lighting, directly or indirectly, such as skin cancer, eyestrain and headache . By studying light ergonomics, we will be able to know how lighting can be an important factor in providing a healthy interior for an individual.

2. Objectives

The objectives of this research are:
i. To understand the basic principles, knowledge, and requirements of interior lighting relating to health and well-being.
ii. To identify the good and poor lighting based on the observation and findings for future design references.
iii. To understand types of lighting existed and their pros and cons.

3. Research Method

This research refers to the previous studies as to support the latest research findings. Qualitative data were collected, each with its own merit in strengthening the study and to support the scenarios, which are further supported by visual evidence. The observation covers 500 shop lot from ground floor to third floor from 11:00 a.m. until 10:00 p.m. Five researcher were focusing on the social interaction that stimulate the shoppers on what has happened and is happening related to the artificial and natural light. Results of the discussion and analysis will be analysed globally, and this will be beneficial for future in-depth studies on light and health in the context of Malaysian domestic health.

4. Functions of Lighting in Build Environment

In ancient time, lighting was provided by fire, sunlight, torches and candles. Lighting was primarily produced for people to lighten up their living places, for religious activities, or to carry out daily activities such as reading, yet it was not served to function for buildings, roadways or any other aesthetical aspects. As time passed by, the technology of artificial lighting has been developing, hence the first electrical lamp (carbon-arc lamp) was introduced by Sir Humphrey Davy, an English chemist. Nowadays, the functions of lighting vary following the various types of lighting features. Lighting is the foundation of our vision, it also serves as the most important aspect of interior spaces. It provides visual comfort to the users, enhances the mood of a space, gives an illumination effect and even ensures users' safety in a dark or hard-to-see environment. Undeniably, lighting plays an important role in our daily life. Furthermore, natural lighting, or daylighting, becomes the hot topic in lighting design for its economical and health issues. The utilization of natural lighting can lead to substantial energy saving. Artificial lighting consumes high amount of electricity, so the use of daylighting can help reduce the energy used for electric/electrical lighting by 20 to 60 percent. According to scientists at the Lighting Research Centre (LRC) in Troy, N.Y., daylighting could increase occupants' productivity and comfort effectively and provide mental and visual stimulation which are necessary in regulating our circadian rhythms [6]. However, excessive daylighting designs will subject users to glaring problems as well as thermal stress. Further related issues will be discussed in this paper. Hence, artificial lighting and daylighting are important in our life. Designers have the responsibility to choose the suitable lighting design to fit the function of a space. One of the most important factors that every designer should consider is the visual comfort of users in order to increase the productivity and their quality of life.

5. Light and Health

Light is important for humans' health. Light, or light radiation, not only affects our visual cortex but also our alertness, wellbeing and performance [2]. Our circadian rhythm and circadian clock are genetically fixed, but they are regulated to a certain extent by our surroundings, such as light. The photoreceptor cells in human eyes- Retina are sensitive towards light. These cells have a nerve connection to the biological clock and the pineal gland located in the brain, which is responsible for the regulation of hormones in our body. In other words, there is a direct relationship between light, body timing and hormones. Cooper and Breiling [4] have listed the benefits of light to human's health and wellbeing as:

1- Light is a "186,000 MPS (300,00 KPS) nutrient.
2- Light has "drug-like-effects" on hormones and neurotransmitter which affects every cell in the body.
3- Bright light can synergize medications in the treatment of the Seasonal Affective Disorder, major depression and Alzheimer's disease.
4- Light "lights's up" people's brains and their lives with both instant and long-term effects.
5- Light and darkness time every major biochemical reaction in our body such as sleep-wake fullness, appetite, mental efficiency, immunity and energy.

Further to this, there are a few bodily processes such as the sleep-wake rhythm, the rhythm in body temperature and the rhythm in which certain body hormones produced are influenced by the light-dark rhythm of the Earth. Biological rhythms that repeat approximately every 24 hours are called circadian rhythms. It is important for us to be exposed to the light and to receive an adequate light dose in order to keep the body clock in sync with the 24-hour day rhythm. Our hormonal secretion is under the influence of the light- dark rhythm. The cortisol, or also known as energy hormone, increases during day time and decreases at night but remains at a sufficiently high level to maintain blood sugar level and alertness. The level of melatonin, or the sleep hormone, decreases during daytime to reduce sleepiness, but rises at night to facilitate sleep. Hence, excessive exposure to the light will lead to circadian disruption which will affect our sleeping quality, physiological functions and neuro-behavioural performance. Moreover, the sensitivity of the photoreceptor cells in human's eyes enable us to make use of lighting for therapeutic purpose especially for those with disturbances in their biological clock Photoreceptor Cells [7]. For examples, therapies for certain forms of sleep disorder, seasonal affective disorder (SAD) and jet lag or sleep-wake rhythm problems [3].
Besides that, the sun which is the main lighting source of the earth, has brought a lot of advantages to humans as well as flora and fauna. The sunlight is necessary for plants to carry out photosynthesis process. Glucose is produced during photosynthesis which then provides energy to the plants. Animals eating the plants will indirectly consume the glucose produced to survive. For humans, the exposure to the ultraviolet-B radiation in the sun rays causes skin to create vitamin D which helps to build strong teeth and skeletal health [8]. The low level of vitamin D can lead to rickets in children and bone diseases such as osteoporosis and osteomalacia. The exposure to the sunlight also increases the level of serotonin which helps in boosting mood. Thus, those who are suffering from depression are suggested to expose themselves to the sunlight as they will be able to absorb more sunlight thus increases the serotonin level in their bodies. However, the excessive exposure to sunlight is harmful to our health and can even become fatal. It can quicken the aging process as well as the reddening of the skin, pigmentation development and skin cancer.

6. Findings and Discussions

The M Mall O2O opened in October 2015, which then became the latest shopping mall present in Georgetown. The interior of the mall was supposedly aimed to provide a new experience to visitors. It contains several attractions: the first wax museum on Penang Island, nationally themed "street" at different levels and various artistic decorations. Once entering the shopping mall, the main walkway is directed by well decorated ceiling. Artificial lighting fixtures are used in ceiling decoration. The types of lighting used are of at different intensities and temperatures. Some of the types of lighting are less bright as to create a relaxing and comforting atmosphere, beautifying the products displayed in the shops. In this case, lighting is important for enhancing the mood of the shops and hence is able to promote the sales. Recessed down lights found is at a slightly poor quality level as most of the time glares are produced. The lighting source used is a mix of light emitting diode (LED) and florescence lamp. The lighting's colour temperatures range from 3000K to 3500K. Furthermore, glares are produced in the shops due to the high intensity of the lighting source. As polished lighting-coloured marble floor finishes are used, the internal reflection index would increase and sometimes would create glares, making the occupants' eyes uncomfortable. Besides, different colours of lighting sources with different wavelengths are used to decorate the interior. It creates a chaotic vibe and would probably irritate the eyes of the occupants (Figure-1).

However, the overall ambiance of the shopping mall on the ground floor is delightful and effectively promotes the mood of visitors. A welcoming effect is successfully established through its lighting. The first floor of the M Mall is a level full with decorative lighting. The ceiling is decorated with artificial lighting fixtures ranging from florescence tube lamps to LED light strips. The walkways are directed by the ceiling design and manage to provide a cosy ambiance for the shoppers. Once entering the first floor, the mood is directed by the decorative ceiling and walls. Translucent acrylics are used to create a story line of fairy tales along the walkway – plastered on the ceiling and the wall. With the aid of the fluorescence tube lamps (colour temperature less than 2700K) which were installed, the aesthetical value of the walkway is enhanced. The lighting and shadow effects stimulate and attract the shoppers, especially the children.

At the same time, at the same level, another space with different ceiling design was found. The ceiling design directed the shoppers along the walkways effectively. Recessed down lights with colour temperature less than 2700K have been installed by the operators. However, some of the parts of the ceiling are tried to be accentuated by the installation of cool white LED light strips. This brings nonsensical clashes on the ceiling. The colour temperature of the LED light strips is rather high (4000K - 5000K) and contains more blue lights which make them look brighter compared to warmer, white LED light strips.

Nevertheless, the head room of the first floor collides with the Universal Building by Law. It is low and is very dangerous for the visitors as they might hit the ceiling if they are taller than others. Moreover, the distance of the lighting fixtures installed are too close with the visitors' eyes. This might lead to the irritation to the eyes and hence might reduce the mood of the shoppers. Glares produced by the excessive artificial lighting fixtures installed may tire the eyes and emotionally affect the visitors (Figure-2). The amount of lighting fixtures used should be reduced in order to give the visitors a better shopping experience. LED light strips may be a good choice but a warmer colour temperature is better to reduce a high contrast. Besides, the usage of fluorescence tube lamps which are being switched on all the time will cause the occupants - the shoppers as well as the workers – to be exposed in excessive UV radiation emission. In a prolonged period of time, the visitors might easily experience fatigue and dizziness. In addition, the themed "streets" placed the shoppers in a Wonderland. There are also fanciful decorative lighting applied to create a reverie ambiance. There are attempts to mix daylighting with lighting installation in the interior spaces, however, the outcome seems to exhibit incongruity. Pendant lights, fluorescence tube lights and recessed down lights are used for the interior. Fluorescence tube lights with different colours are installed, emphasizing the big decorative elements across the ceiling, at the same time guiding the shoppers. The random light colour used for the lighting fixtures looks orderless and this might make the shoppers confused with its purposes.

Daylighting, which is good for health, is introduced through the big floor to ceiling windows. However, the improper filtration of daylighting leads to failure as it is unable blend well with the artificial lighting installed. One of the shop-selling sport equipment was identified to display this problem.

Figure-1. Glare in shopping mall from the daylight

Figure-2. High intensity of lighting level

During daytime, the recessed down lights are switched on although it is functionless in terms of enhancing the brightness of the interior and this activity consumes a lot of energy. In this case, the recessed down light becomes a waste during daytime. The light should be only switched on when there is no daylighting or if the daylighting intensity is too low. The spacious floor and broad ceiling windows in the shop should be able to bring in sufficient amount of daylighting to the interior during daytime as well as enable the workers to carry out their professional routine and the shoppers to enjoy their shopping. Furthermore, tracking lights may be switched on to promote the products by attracting shoppers. With the lighting and shadow effects created, the products will look interesting and will be able to give more excitement and enjoyment of shopping within consumers.
The researcher therefore has synthesized that the phenomenon of artificial and natural light at M Mall has triggered customer's emotion and visual perception through (Table-1):

Table 1. Phenomenon of artificial and daylight in M Mall

S. No.	Issues
1	Design tools & techniques (high intensity, high contrast, directionality, colour).
2	Excessive color light by making an impression of the ceilings and architectural elements.
3	Enhancing interior spaces by highlighting columns, ceilings, walls and decorative elements.
4	Creating drama through interior themes using colour light.
5	Great impact on the visual and eyes from the general lighting of the fresco and decorative ceilings
6	Improper filtration of daylighting

7. Conclusion

In conclusion, through this case study, designers should be able to understand what lighting is about and how it can affect people's health. Designers should understand ways to exhibit good lighting for interior space. The placements and types are important aspects of interior design as they work together with colour sections, space, and furniture selection. Depending on the types of the bulb, the light emitted may be of a warm or cold colour, and this may affect the overall feeling of a room. Good designers should not be selfish by considering just the aesthetical factor, for an instance, by placing light everywhere around the town and rural areas without consideration. In terms of the health aspect, the effects of light on human health as well as on the other living things need to be prioritized. Lastly, without proper lighting, interior spaces cannot be experienced the fullest. Good lighting assures warm, inviting yet functional interiors.

References

[1] Boivin DB, Czeisler CA, et. Al (1996), "Dose-Response Relationships for Resetting of Human Circadian Clock by Light," Nature 379:540-542.
[2] Bommel, Wout van. (2005). Biological effect of Lighting. ELDA + Light Focus 2005. 14-15 April 2005 in Milan, p. 5.
[3] Circadian Rhythms Fact Sheet. https://www.nigms.nih.gov/Education/Pages/Factsheet_CircadianRhythms.aspx accessed on 22 February 2017.
[4] Cooper, P. and Breiling, B., Light Meditation and Introductive Overview. PLDC 1st Global Lighting Design Conference, 24-27 October 2007 in London, p. 30.
[5] Funk, Rischard. Benefit and risks of light entering the eye. PLDC 1st Global Design Conference, 24-27, October, 2007 in London, United Kingdom.
[6] Lighting Research Center. http://www.lrc.rpi.edu/ accessed on 22 February 2016.
[7] Photoreceptor Cells (Photosensitive Cells). https://www.ncbi.nlm.nih.gov/pubmedhealth/PMHT0024257/ accessed on 22 February 2017.
[8] Sunlight and Vitamin D: A Global Perspective for Health. https://www.ncbi.nlm.nih.gov/pmc/articles/PMC3897598/ accesses on 22 February 2017.

IOP Conf. Series: Materials Science and Engineering **291** (2017) 012006 doi:10.1088/1757-899X/291/1/012006

Case study on perspicacity of collaborative learning experiences

Fadzidah Abdullah[1], Noor Hanita Abdul Majid[1], Ibrahim Numen[2], Aida Kesuma Azmin[1], Zaiton Abd. Rahim[1], Zuraini Denan[1], and Muhammet Emin Sisman[2]

[1]Department of Architecture, International Islamic University Malaysia, Malaysia.
[2] Department of Architecture, Fatih Sultan Mehmet Waqf University, Istanbul, Turkey

fadzidah@iium.edu.my

Abstract. In the attempt to relate to the architectural practice, architectural education today has augmented the development of collaborative learning environment in the campus scenario. Presently, collaborative work among students from the same program and university is considered common. Hence, attempts of collaboration is extended into having learning and teaching collaboration by means of inter-universities. The School of Architecture, at the International Islamic University Malaysia (IIUM) has explored into having collaboration across the continent with Fatih Sultan Mehmet Waqf University (FSMWU), among faculty members and students of the two (2) universities This paper explicates the empirical study on students' perspicacity of their collaborative learning experiences; in term of effectiveness, generative behaviour, and teamwork. Survey with three (3) open-ended questions are distributed to students to express their opinions on learning collaboration that they have had during the execution of the Joint Summer School Program (JSSP). Feedback on their perspicacity is obtained and organised into numerical and understandable data display, using qualitative data processing software. Albeit the relevancy of collaborative learning, students gave both positive and negative feedbacks on their experiences. Suggestions are given to enhance the quality of collaborative learning experience for future development

1. Introduction

Recently, collaboration is considered a vital method of skill development in order to create leadership and teamwork skills among graduates. Architectural education has also been impulsive to respond to new sets of collaborative requirements, and has augmented the envelopment of collaborative development in learning environment in campus scenario. Most schools of architecture throughout the globe have initiated multi-fold collaborative educational programs, where numerous forms of collaboration between numerous parties are established

Similar collaborative program has been organised by the School of Architecture at the International Islamic University Malaysia (IIUM) and Faith Sultan Mehmet Waqf University (FSMWU), to provide a unique collaboration for teaching and learning in a program called the Joint Summer School Program (JSSP). JSSP is intended to instil collaborative skills among Malaysian and Turkish architecture students, as well as among the academic staff from both countries. This program intends to offer new opportunities for students to explore international collaborative learning environment, and paves new horizons in students' recognition to different opinions, views, philosophies, and cultures. Both universities agreed on having architectural design studio as the platform to acquaint students with

Content from this work may be used under the terms of the Creative Commons Attribution 3.0 licence. Any further distribution of this work must maintain attribution to the author(s) and the title of the work, journal citation and DOI.
Published under licence by IOP Publishing Ltd

diversity of learning methods and processes, and simultaneously strengthen the teaching collaboration between the two (2) universities.

2. Objective and methods

This case study research has the objective to explicate students' perspicacity of their collaborative learning experience; in-term of effectiveness, generative behaviour, and teamwork. Survey with three (3) open-ended questions were distributed to students to express their opinions on learning collaboration that they have had during the execution of The Joint Summer School Program (JSSP) between architecture students of IIUM and FSMWU. Participants from IIUM are twenty (20) 3rd year architecture students, whilst eighteen (18) students from the turkey counter-part are from the second, third, and forth year architecture students.

The first question requests students to express their opinion on the effectiveness of group work in design studio. The second question probes if the team members have valuable or detrimental behaviours during the collaborative learning process, and the third question seeks to examine lessons learnt to be implemented for future collaborative projects.Feedback on their perspicacity are obtained and organised into numerical and understandable data display, using qualitative data processing software. The findings are descriptive and would be presented in graphic form for ease of understanding.

3. Collaboration in architectural education

Collaboration is supposed to be the strength of the architectural profession[1], and architectural education is often blamed for failing to produce good architects with teamwork and communication skills[2]. Key skills for architectural professional practice; such as listening to others, questioning, and negotiation; are generally not developed in the undergraduate years[3].

In the attempts to encourage collaboration in architectural education, Lehmann[4] highlights three (3) developed models with varying levels of cross-collaboration. Collaborative Studio model 1 is "Trans disciplinary Encounter Model", where students work side-by-side on the same project, each student would be producing their own proposal, whereby the contributing disciplines remain clearly identifiable. Model 2 is "Real team Collaboration," where students working in multidisciplinary teams, encompassing architecture and visual arts students, to collaborate on joint individual projects. The third is "Interdisciplinary Consultation Model", in which students invite other students from other disciplines to come to the studio from time to time as external consulting experts. This model is probably closest to the standards of "real" architectural practice, operating with "informed disciplinarily', calling upon and utilizing the expertise of other disciplines. Hence, collaborative learning is useful to develop the long-life learning capability among architecture students.

On the other hand, modern development of world-wide-web (www) and the internet of things (IOT) have cultivated new measures of collaboration in architectural education[5]. To remain relevant, architectural education must adapt technology enhancement collaborative learning, and ensure education framework are to be aligned with the current world.

4. Research Analysis

This research quantifies qualitative data obtained from open-ended questionnaires. Survey were distributed to architecture students who participates in the JSSP. With different background and cultures of students, it is expected that upon analysing students' opinions, several concerning matters are to be highlighted for future program to be implemented.

20 Malaysian students and 18 Turkish students participated in the organised JSSP 2016 program. From the 38 participants, 34 students responded to the distributed survey (89.4%). Based on 34 collected questionnaires, there are 117 quotations in all four main codes which are grouped as effectiveness (25 quotations), team members behaviour (24 quotations), a) behaviour valuable in the group (15 quotations), b) behaviour detrimental in the group (9 quotations), lesson learnt during the group work (35 quotations) and concerning matters during the group project (30 quotations). Table 1 shows the description and statistics summary of every analysed codes and quotation.

Table 1. Analysed codes and quotations

Codes No	Codes Name	Freq.	(%)
1	**Group effectiveness**	**25**	-
1.1	• Very effective	16	64.0
1.2	• Effective	8	32.0
1.3	• Not effective	1	4.0
	Total	25	100
2	**Team members behaviour**	**24**	-
2.1	*Behaviour -valuable*	15	62.5
2.1.1	• Team work spirit	8	33.3
2.1.2	• Helpful	1	4.17
2.1.3	• Active in participation and hardworking	3	12.5
2.1.4	• Knowledgeable and skillful	2	8.33
2.1.5	• Very fluent in speaking and translating	1	4.17
2.2	*Behaviour- detrimental*	**9**	37.5
2.2.1	• Disturbing group mates during working	1	4.17
2.2.2	• Reluctant to do assignment/tasks	1	4.17
2.2.3	• Reluctant to spend more time to do works	1	4.17
2.2.4	• Not appreciating group mate's ideas	2	8.33
2.2.5	• Underestimate group mates' capability	2	8.33
2.2.6	• Low working motivation	1	4.17
2.2.7	• Not communicate with other group mates	1	4.17
	Total	24	100.0

Codes No	Codes Name	Freq.	(%)
3	**Lesson learnt that can be used for the next group work**	**35**	-
3.1	• Collaborating in team work	3	8.57
3.2	• Sharing, choosing and delivering ideas	3	8.57
3.3	• Respects different ideas	3	8.57
3.4	• Learning different architectural systems, design concepts, design process and design methods	6	17.14
3.5	• Delegate job effectively	2	5.71
3.6	• Learn leadership and team management	5	14.29
3.7	• Communication and presentation styles	7	20.00
3.8	• Preparation for future career	2	5.71
3.9	• Using advance architecture software like Revit	4	11.43
	Total	35	100.0
4	**Concerning matters**	30	-
4.1	• Communication-English ineloquent and dialect	22	73.33
4.2	• Do not learn and use advance architecture software for assignment project/ university syllabus.	3	10.00
4.3	• Different level of understanding, different process to proceed more in designing	2	6.67
4.4	• Cultural difference or adaptability	1	3.33
4.5	• Hard to gather group member	1	3.33
4.6	• Crowded environment	1	3.33
	Total	30	100.0
	GRAND TOTAL	**117**	

5. Survey Result

Overall, for the group effectiveness score, about two-third (64%) of the students rated the group work as very effective to them, about one-third (32%) rated the group work as effective and a small number (4%) rated the group work as ineffective to them. The group work effectiveness was evaluated based on their responses on how the group work influence their personal view toward their own group because literally they do not response specifically according to the survey question. Figure-1 shows the result of analysis on effectiveness.

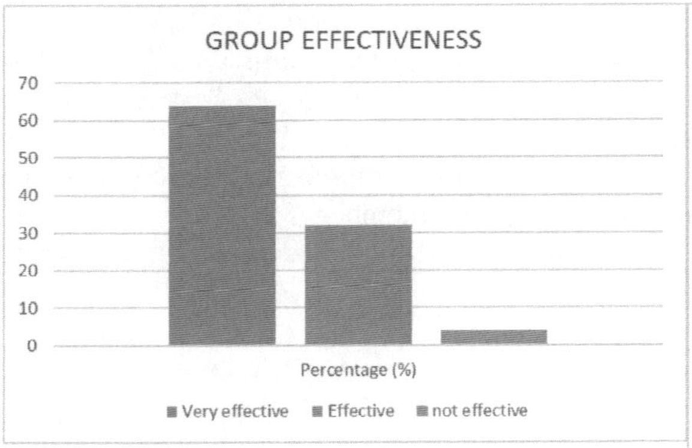

Figure 1. Result of analysis on effectiveness

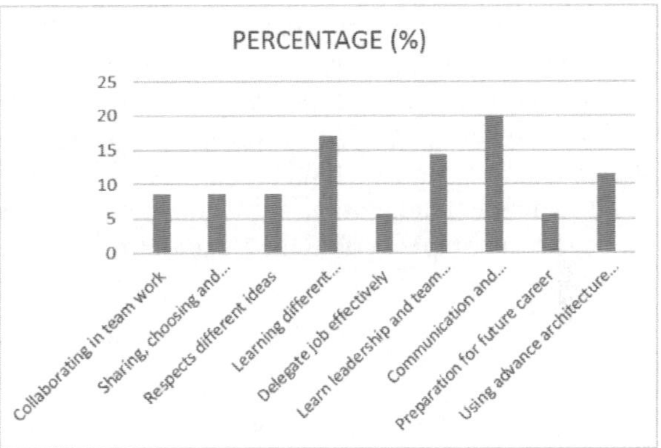

Figure 2. Lesson Learnt

For the team members' behaviours, the findings revealed that the valuable behaviour that were recorded, were higher (62.5%) than the detrimental behaviour (37.5%) during their group. Among the valuable behaviours highlighted are, Team work spirit (33.3%); Helpful (4.17%); Active in participation and hardworking (12.5%); Knowledgeable and skilful (8.33%); and Very fluent in speaking and translating (4.17%).

The students, both IIUM and FSMWU, highlighted that their group mates have showed positive team work spirit (33.3%), active as well as hardworking (12.5%) and knowledgeable (8.33%). Other students response appreciation towards their friends' helpful acts (4.17%) and very fluent in speaking (4.17%), thus reduce the communication gap (refers to the concerning matters at the end of the report) during the group works.

Whereas, the detrimental behaviour recorded are as follows, disturbing group mates during working (4.17%); reluctant to do assignment/tasks (4.17%); reluctant to spend more time to do works (4.17%); not appreciating group mate's ideas (8.33%); underestimate group mates' capability (8.33%); low working motivation (4.17%); and not communicate with other group mates (4.17%).

For the detrimental behaviour findings, some of the FSMWU students reported that the Malaysian students (IIUM) were not appreciating their ideas in making group decision. Some of the IIUM students reported that students from FSMWU underestimates their group mates (IIUM) capability related to the application of advance architectural computer software.

The students highlighted numbers of responses for the question of lesson learnt that can be used for the next implementation. Among the highest rated lessons learnt, the students highlighted they have learnt useful communication as well presentation styles (20.0%), different architectural systems (17.14%), leadership together with team management (14.29%) and using advanced architecture computer software (11.43%). Other useful lesson learnt for their next project were listed as Collaborating in team work (8.57%); Sharing, choosing and delivering ideas (8.57%); Respects different ideas (8.57%); Delegate job effectively (5.17%); and Preparation for future career (5.71%). Figure-2 shows the percentage of opinions on the lesson learnt from students' experience that could be applied to future implementation.

For the last section, among the concerning matters raised by students are as follows: the majority of the students (73.33%) highlighted the issue of English ineloquence as the main reasons to deter communication and group discussion. Some of the students (10.00%), particularly the IIUM students emphasised on the lack of course provision in advanced architecture software provided by the department.

Some of the IIUM students commented that most of the FSMWU students were very advanced in architecture software such as Revit. However, some of the FSMWU students were also focusing more on manual techniques in their studio, thus they face difficulty in adapting to the new application to design. This finding suggests the necessity of having specific architectural software workshop or class in relation to the industrial practice. One minor issues raised by the students are having opposing perspectives in design process (6.67%), coming from two different schools of architecture. Some of the students also review cultural difference (3.33%) as the reason of miscommunication during the group work, which could be associated with another issue to unite the group member (3.33%). Also, some of the students complained on the issue of crowded environment (3.33%) which had caused discomfort to the group work.

6. Conclusion

Hitherto, this research has investigated the students' matters of concerns during the Joint summer school program between IIUM (International Islamic University Malaysia) and FSMWU (Fatih Sultan Mehmet Waqf University) and students' perspicacity on the learning collaboration they have experienced in the joint summer school program (JSSP). The findings were derived from the qualitative data analysis software and translated into statistics figures for clearer findings.

This research has revealed that the joint summer school program has effectively generates group work spirit among students, as only 1 out of 38 students found the program was ineffective. The program has also shown that there are positive and negative generative behaviours that students had during the process of learning collaboratively. However, in-term of percentage, valuable behaviour has surpassed decremental behaviour during the learning process; thus, it could be concluded that the minimal decremental behaviour could not jeopardise the whole group work performance.

Students have also expressed their opinions on numerous lessons that they had learnt, which could serve as important aspects to be considered for future implementation of the same program. The most important lesson highlighted is, they learnt to understand inter-cultural differences; in-term of communication, design presentation styles, leadership, and team management.

In fact, communication is perceived as the most dominant concerning matter that students expressed. Although English is the main medium of communication, but both groups of students do not come from English speaking nations. Thus, they speak English with their folkloric dialects that makes understanding each other difficult. Therefore, for future program, similar level of English eloquence among students should be one major aspect to be considered for future implementation of joint summer school program. In general, the information generated from this research should provide insight for the department in hosting an efficient joint program with other university.

References

[1] Parnell R 2003 Knowledge skills and arrogance: educating for collaborative practice. *EAAE Transaction on architectural education* **15** Copenhagen: EAAE.

[2] Egan J 1998 *Rethinking construction - the report of the Construction Task Force*. London: RIBA.

[3] Nicole D and Simon P 2000 *Changing architectural education: towards a new professionalism*

IOP Conf. Series: Materials Science and Engineering **291** (2017) 012006 doi:10.1088/1757-899X/291/1/012006

Spoon Press London

[4] Lehmann S 2009 Interdisciplinary models for collaboration between artists and architects: Empowering community, inspiring urban renewal. *UNESCO Observatory E-journal* **1** 4

[5] Traxler J et. al. 2016 *Mobile learning: The Next Generation* New York: Routledge Taylor and Francis Group

The Use of Computer Tools in the Design Process of Students' Architectural Projects. Case Studies in Algeria

Ouafa Saighi[1] **and Mohamed Salah Zerouala**[2]
[1] Laboratoire "Architecture, Ville, Métiers et Formation", Département de Management de Projets, Faculté d'architecture et d'urbanisme, Université de Constantine 3, Algeria,
ouafa.saighi@univ-constantine3.dz
[2] Ecole Polytechnique d'Architecture et d'Urbanisme, Laboratoire "Ville, Architecture et Patrimoine", Alger, Algeria,

ms.zerouala@epau-alger.edu.dz

Abstract. This The paper particularly deals with the way in which computer tools are used by students in their design studio's projects. Four institutions of architecture education in Algeria are considered as a case study to evaluate the impact of such tools on student design process. This aims to inspect in depth such use, to sort out its advantages and shortcomings in order to suggest some solutions. A field survey was undertaken on a sample of students and their teachers at the same institutions. The analysed results mainly show that computer tools are highly focusing on improving the quality of drawings representation and images seeking observers' satisfaction hence influencing their decision. Some teachers are not very keen to overuse the computer during the design phase; they prefer the "traditional" approach. This is the present situation that Algerian university is facing which leads to conflict and disagreement between students and teachers. Meanwhile, there was no doubt that computer tools have effectively contributed to improve the competitive level among students.

Keywords: Computer tools, Architecture, Design process, Survey

1. Introduction

Since the late 1990s and until recently, students hesitated between buying tracing paper or personal computers. On the other hand, some students (for some other reasons) find it hard to express themselves and make plainly their ideas through paper. So, they would use the computer to make the virtual image speaks on their behalf. This has in fact led to a well-known situation among students which made them dependent on their data bank and/or e-libraries.

As far as teachers are concerned, two diverging behaviors can be distinguished. There are those who believe in the valuable and necessary use of computer tools during the design process, and those who prefer the pencil and the traditional method named by Sabine Porada "the paper architecture" [1]. From the latter, some might go further up to forbid their students to rely on computer, especially in the design stage, arguing that computer inhibits personal creation. The trend of such tools at schools/departments of architecture in Algeria has seemingly led to a dilemma among both students and teachers. However, such conflict can only be transitional and will likely disappear soon.

For reminding, at universities, computer courses are introduced very late (the fourth year: a year before graduation) in the architectural curriculum. This is worsened by the very low credit hour allotted to computing (5,58% out the total weekly credit hour).

2. Methodology

To better grasp the present situation, a first survey was carried out during two consecutive academic years: 2008-2009 and 2009-2010 at the department of architecture, university of Constantine. On 2011-2012, a second field study has covered four institutions (An independent school and three departments at universities: Epau-Algiers, Department of architecture-Constantine, Department of architecture-Oran, Department of architecture-Biskra). The aims were to make a horizontal comparison (within the same year), a vertical one (between different years) and assess any eventual evolution. The questions mainly dealt with the use of computer tools by students within their projects during the different years. In addition, students' preferences (in relation to computer use) were also tested. The questionnaires were distributed to students of third, fourth, and fifth years, as these years are the more oriented towards the core program. The first field study carried out at Constantine has been more detailed in a previous paper [2] based on a sample of 290 students. While the sample of the second field study encompasses 254

Content from this work may be used under the terms of the Creative Commons Attribution 3.0 licence. Any further distribution of this work must maintain attribution to the author(s) and the title of the work, journal citation and DOI.
Published under licence by IOP Publishing Ltd

students. Another survey was conducted with the teachers of the targeted groups, collecting their opinions on the same issue.

3. Results And Discussions

Through our teaching experience and involvement within different architectural agencies, we have been able to witness such a situation, which gave us the opportunity to share with teachers and students the issue about the role and the contribution of computer tools in the project's process. Some of the results are detailed in what follows.

Students

The results of the field survey (2011-2012) show that students calling for computer aid to their studio projects exceed at least 65% by year level. The statistical analysis shows a significant correlation between the two variables: the use of computer tools and the year level (sig<0,05, Chi-square=36,315). Indeed, by mean of comparison between the three levels, positive answers (yes) increased by 31,07% from the 3rd year to the 5th one "Figure 1". This is due to the objectives of the program content set for each year level; for the third year, where students are asked to design several architectural projects, third (1/3) of theme do not use computer.

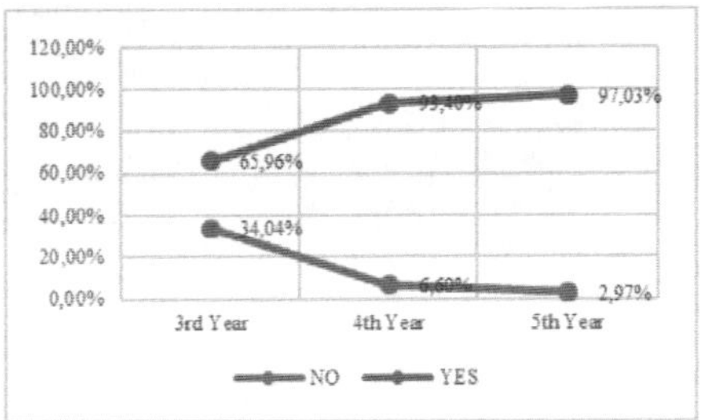

Percentage for each year level

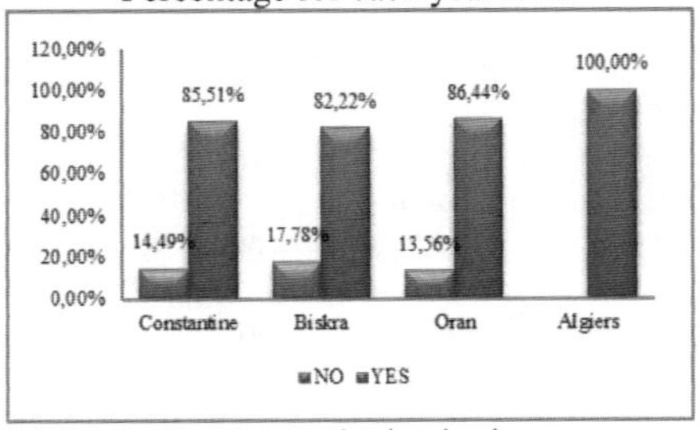

Percentage by institution

Figure 1. Personal use of computer tools by students in their projects during the current year. Authors: field work, May-June 2012.

The four institutions concerned by the survey show close figures. The average percentage of computer use for all institutions is around 88,54%. However, Algiers is the only school where we notice a total use of computer by students regardless the year of the cursus. For the department of Constantine which benefited from the first field study, the use of computer tools has risen by 38,97% through the three academic years.

These high percentages do not mean that students carry out their works without help. The more they progress in their studies, the more contact they make with architectural agencies asking for different services "Figure 2". Statistical analysis shows a significant correlation between the contact of architectural agencies and the year level (sig<0,05, Chi-square=8,673). By this contact, students want to compensate some deficiencies; they know that "beauty/esthetic" has a crucial role for project approval. They tend to impress their tutors and the examiners board through high-quality printing, nice colors and attractive virtual tours. However, the percentage of answers with "yes" which is 32,28% (between the three level years and the four institutions) is misleading. It represents projects carried out outside the university. The observed percentage of Constantine department for the same academic year (2011-2012) is higher (43,48%), it evolved by comparing it to the results obtained in the previous academic years. During the first field study, it reached 34,39% [2]. The school of Algiers seems to be the least "tolerant" concerning the contact of architectural agencies (among the four institutions), with the lowest percentage (17,28%). At the 3rd year level, no student from Algiers has contacted any architectural agency. This can be explained according to our investigation by the firm evaluation, as the juries' members are usually external to the institution.

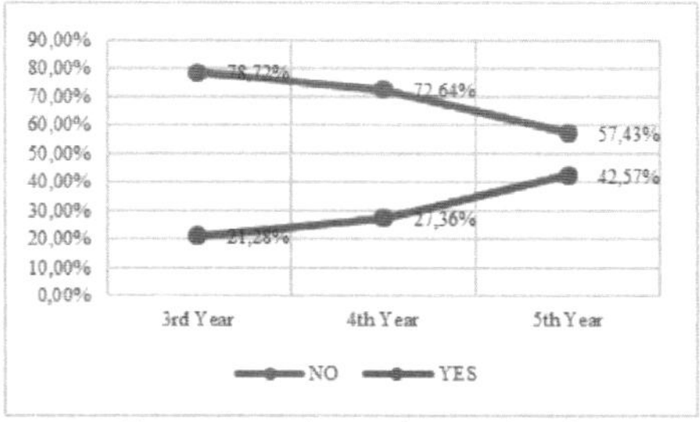

Percentage for each year level

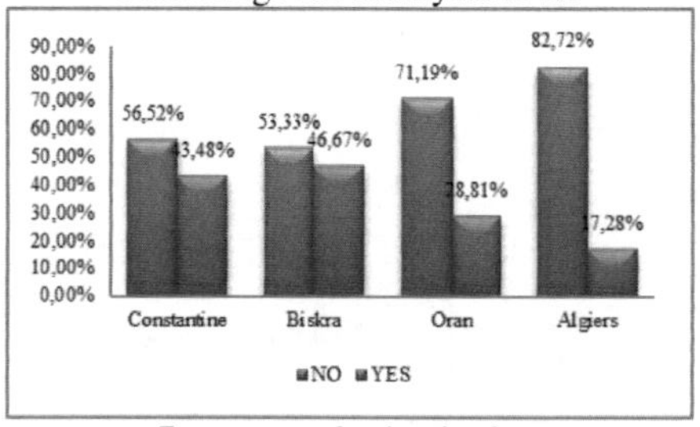

Percentage by institution

Figure 2. Contact of architectural agencies for different tasks for students' projects.
Authors: field work, May-June 2012.

Despite the objectives set for each year, a small percentage (15,44%) is noticed showing students' contacts with architectural agencies seeking help during their first year of studies, although the first-year training objectives concentrate on mastering free hand drawing. We wonder then upon students' attitudes and teacher's requirements. Teachers certify through the survey, that they are sometimes surprised -the day of the exhibition- about their students' projects that clearly show a part of the product carried out outside the university. This leads to an evaluation problem putting the staff in subjective situation. They are aware that students should undertake each project under their own supervision until the final exhibition. The difficult mission lies on the fifth year' students (final step of the curriculum). They benefit from a large time before projects' exhibition. Most of it seems to be used to polish the quality of the work, with the help of the architectural agencies.

For this reason, students need to spend money that (the more often) exceeds 450 euros for each project. 27,36% of them spend for one project more than the average of their annual grant. These expenditures increase from one year to the next. Indeed, the Spearman test shows us a significant correlation between the spent amount and the year level; the more students progress in their studies, the more they spend (sig<0,05, the correlation coefficient = 0,3).

Such services (digitalization and finishing the incomplete plans, 3D models, rendering and virtual tours) has become a common and fruitful business. The survey demonstrates that 18,65% of students contact architectural agencies mostly for rendering their works as we noticed from the first interpretation of the results. In a second sort, all "other" answers to this question gathered from the questionnaires have been categorized, as well the percentages have been recalculated. We were surprised to find that 25,81% of students who contacted the architectural agencies did it for drawing. Their concern, then, does not lie on the image of the project, but it is rather located at the standard of drawing (ability to draw). Hence, questions may arise; we wonder whether the student cannot achieve his work "alone" or the teachers' requirements are beyond his ability to draw.

These students bring their laptops to classrooms and studios to discuss their projects with their teachers. The field survey has showed that the percentage of students who had a training or trained themselves in CAD/BIM raised at the department of Constantine from 73,89% during the academic year 2008-2009 to

86,96% during 2011-2012. This is actually not far from the percentage average (75,59%) of the four institutions. The percentage increases as students pass from one year to another. The more the student advances in his studies, the more interest he develops towards this specific training. The analysis confirms a significant correlation between the training and the year level (sig<0,05, Chi-square = 6,059).

Should we mention, that from the trained students, only 22,04% of them had their training at the university. The remaining percentage call upon self-training and a training that took place outside the university. This could have a bad side effect on teachers' evaluation upon real students know how (competences) gained outside the classroom. But whatever the place, students in Algeria struggle to train themselves in the use of software to improve their skills during their studies in order to prepare themselves for their future professional career. They are fully aware that mastering computer tools is a criterion to get a job. This was confirmed through our interviews with agencies' managers.

Students make use of a rich range of specific software in their projects. Majority of students (93,75%) seem familiar makes use of AutoCad software. SketchUp comes in second position with 60,94%. The dominant situation of AutoCad is due to its wide adoption by all administrations, architectural agencies and firms. At Constantine, it is a widespread software since the early 1990s. At that time, a public company - dealing with study and urban construction- purchased this software and trained its employees to use it. In fact, AutoCad is the favorite software used by students, since it is suitable for drawing and representing. It is well confirmed through students' choices detailed in the title 3.3. Actually, the dominant use of AutoCad is our first indicator showing that the majority of students does not use the computer and its tools to be helped in designing.

Teachers

A glimpse towards teachers shows that some of them do not allow the use of computer as we said before. We asked students if they were "authorized" to use such tools. The results obtained were as follow:

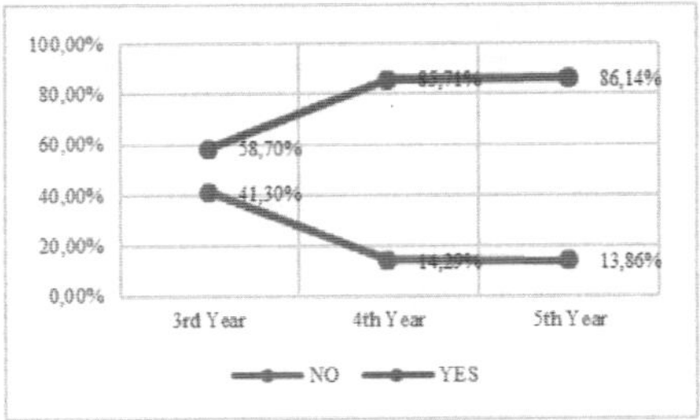

Percentage for each year level

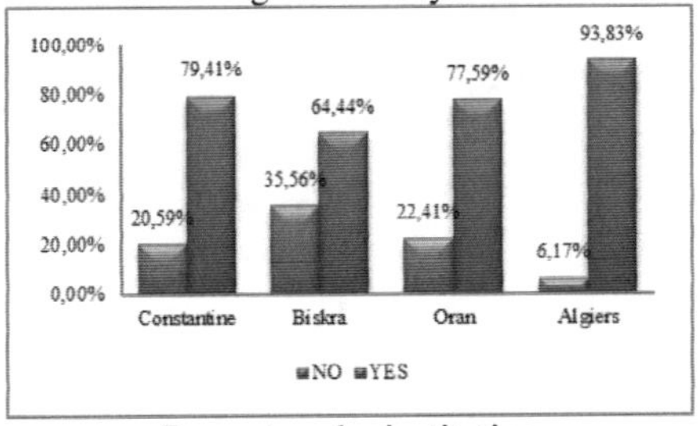

Percentage by institution

Figure-3. "Authorization" to use computer by teachers.
Authors: field work, May-June 2012.

The questionnaires submitted during 2008-2009 showed that teachers of the fourth-year level are divided into two groups (at nearly equal percentage) [2]: those "authorizing" the use of computer tools, and those "forbidding" it and arguing that the objectives of this year level did not match with the use of such tools. Few years later, the situation has changed. The second field study (2011-2012) reveals that teachers are more "tolerant" towards the use of computer; 80,95% of interviewed students appeared to be affirmative. The answers by No in the 3rd year level are higher: 41,30% which is justified by the pedagogical aims of the year. This is well seen through the results showed in the "Figure 1". On the other hand, teachers of Algiers are the most "tolerant" among the four institutions, with the lowest average of NO equal to 6,17%, confirming the widespread use of computers by students "Figure 1". Consequently, "Figure 3" confirms the results showed in the "Figure 1"; the answers for the use of computer tools through the three years matches proportionally with the teachers' agreement. The statistical test confirms this point; a positive correlation exists between the use of computer tools and their teachers' "authorization" (sig<0,05, khi-square = 62,977)

These teachers do not deny the advantages of the computer for graphic presentation, editing possibilities, printing with different scales, but they blame students for using it in the design phase, which can eliminate individual creativity. They claim -as what collected- the lack of sensitivity: the architectural project has become an object reduced to its technical and functional character, putting aside the artistic and symbolic dimension. That, in fact, what Gehry calls: the missing link when evaluating an architect's work [3]

Teachers' point of view and their antagonist positions are backed up by the lack of a clear educational policy at universities. Besides, there are no rigorous rules dealing with the use of computer tools through students' works, although architects' attitudes are erratic. There are those who would still prefer the use of the pencil and free hand sketches, considering the computer tools as a burden and an obstacle at early stages of designing (this issue has been developed and solutions were suggested through Mascaro's paper [4]). And there are those who believe that the computer pushes "architects to become architects more than any time before" [5] as it sustains the process of the project and highlights its strategies.

The teachers' opinion concerning the use of computer tools in the process of an architectural project has been collected through their own survey. The results show that 65,22% of teachers leave the initiative to their students. Only the third, affirms that the use is suggested or imposed by the pedagogical approach. Other teachers have preferred to precise that the computer tools are essential. According to the survey, 60,78% of the latter are familiar with CAD-BIM.

Using computer tools in the process of students' projects

Students were asked which tool they use at every step of their project, especially for representing and designing. Computer tools (used alone) were found as the first choice to **represent** projects: 63,78% for the academic year 2011-2012. High percentage is prevailing from one year level to another and in all institutions. Unlike the use of the pencil and hand drawing which decreases from one year to another "Figure 4".

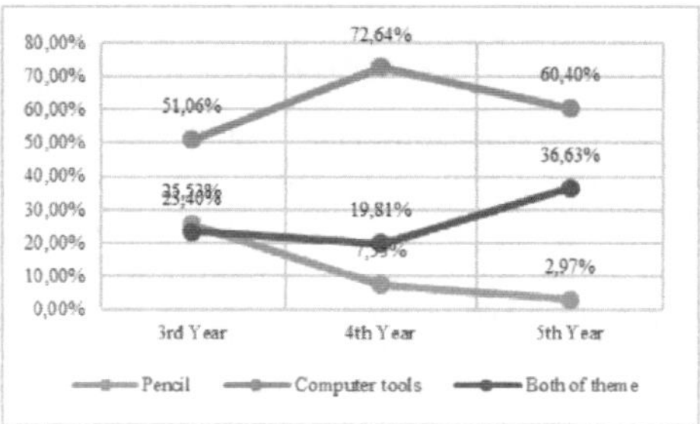

Percentage for each year level

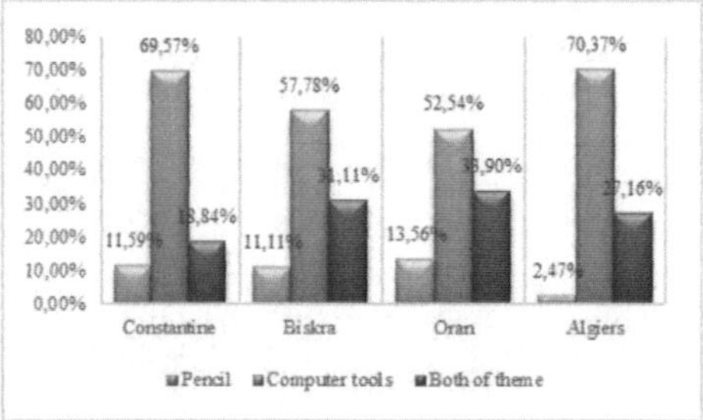

Percentage by institution

Figure 4. Students' choice for representing.

IOP Conf. Series: Materials Science and Engineering **291** (2017) 012007 doi:10.1088/1757-899X/291/1/012007

Authors: field work, May-June 2012.

For **designing**, the pencil (used alone) and the use of both of tools "are jostling" to have the highest percentage during the three years according to the students' choice (44,49% for the first tool and 47,64% for the second one). The use of computer tools "alone" seems to be insignificant regardless to the year level or the institution (7,87% as an average) "Figure 5"

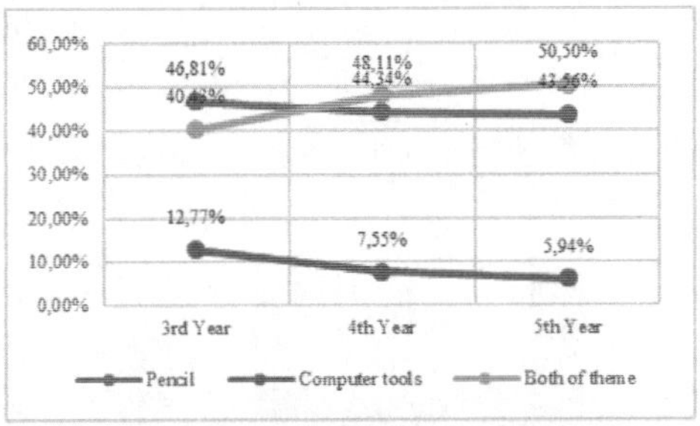

Percentage for each year level

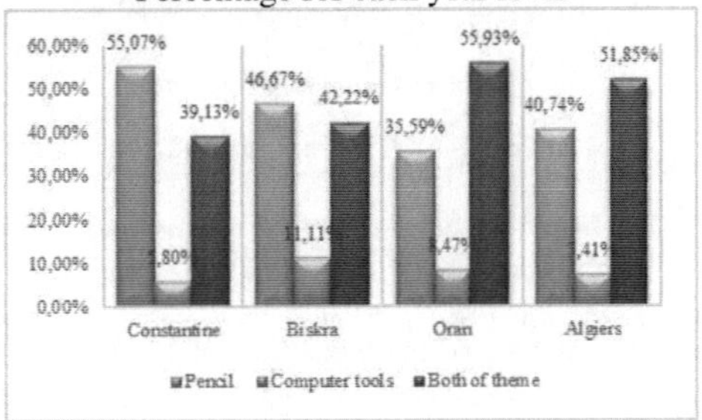

Percentage by institution

Figure 5. Students' choice for designing.
Authors: field work, May-June 2012.

As results of our observations and surveys, the majority of students use computer as an alternative for sketch board and the paper in order to digitalize projects produced in the traditional way. Once the work on paper is finished, students go for digitalizing it using the computer and specific software. Thus, in most cases, the computer and software are used as a tool of drawing. Unfortunately, till our latest survey, computer seemed not be well integrated in the design process (genesis of the main idea, forms, search of solutions and simulation of projects...). Although small progress is noticeable. Some students believe that the computer tools can help in designing but with the help of the pencil, as proved by their explanations to the choice "both of tools", eg.:

- Pencil deals first with primary ideas while improving them rely on computer.
- Free hand sketches in 2D are transformed to 3D owing to computer tools before choosing the final model.

This attitude is, in fact, confirmed by the results showed in the "Figure 6". Through a direct question addressed to students about their first reaction towards the design of projects, the majority of students preferred free hand drawing as a first step. A high percentage exceeding the two-thirds in each year level (at least), expresses such a choice. As an average between levels and institutions, the percentage reaches 90,12%. Students' arguments rely on the fact that the pencil is the basic tool for architectural design and inspiration. It allows first ideas to freely emerge while the computer could not draw ideas. Some students added that the pencil gives the feeling that project belongs to them; once ideas are clearly formulated, computer is left for the last step of the design process.

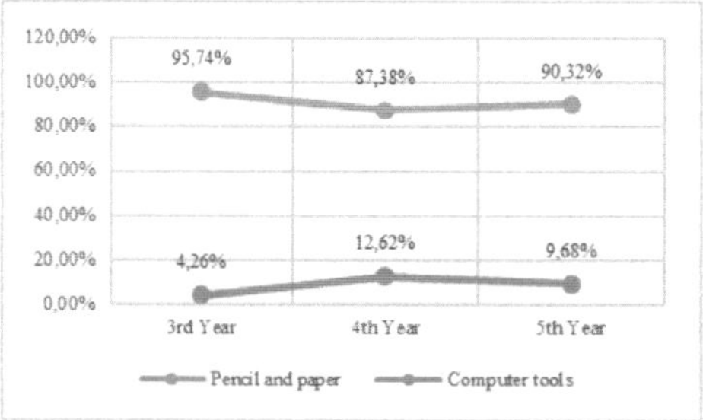

Percentage for each year level

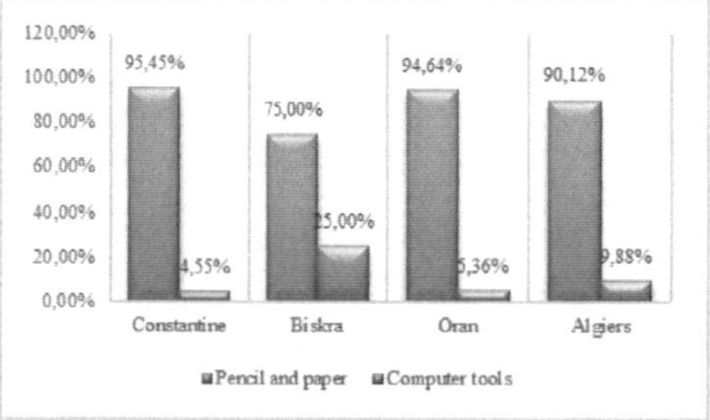

Percentage by institution

Figure 6. Students' first reaction.
Authors: field work, May-June 2012.

However, the majority of those who answered the question dealing with the first reaction for designing insist that projects are carried out back-and-forth between the paper and the computer tools. Any changes are then brought by hand on printed plans. Students and teachers are still insisting on the use of the hand for exploring and modifying suggested layouts. This is, in fact, an "old" behavior which sticks that computer can never replace the hand and the tracing paper, which are essential for searching and correcting [6].

Generally speaking, the project process appears to be totally accomplished with computer tools. The different steps consist of: designing, representing and communicating the project to the others (without mentioning the construction and manufacturing). But, among the four study institutions, the survey confirms that we're still talking only about the two latest phases (representing and communicating) while the first one is not reached yet. Although we cannot dissociate these phases one from one another; according to Leandro Madrazo, the real effect of computer on architecture starts when there is no more separation between design and representation [7].

Historically speaking and since the emergence of information technologies in the world of architecture, Alireza Razavi writing in 2008, relied on three eras: The first one goes back to 30 years ago, the second to 15 years ago, which is characterized by the discovery of the great potentials of rendering and simulation owing to computer. The current phase (according to his paper) is characterized by a real partnership with the computer which greatly contributed to the development of digital design [8]. However, in Algerian universities, computer tools (as we cannot talk about digital technologies) have not been yet developed to help in the design. We are still living the second phase, dazzled and confused between the tool and the objective (the project). This is the period of hesitation and confusion as we described it at the beginning of our research.

4. CONCLUSION

Owing to LMD reform (License, Master, Doctorate), Algerian university is reaching new perspectives. Thus, all institutions encroach for this new educational system. It has significantly integrated computer in teaching, since this topic is taught in every year level with appropriate objectives. Institutions are keen in including more optional courses for students wishing to improve their knowledge in this issue. We also hope to offer to Master students some workshops specialized in computing and its applications on building, as it is performed, for example, at Media x Design Lab at the Swiss Federal institute of Technology in Lausanne [9]. The latest initiative of the department of Oran (Algeria) that offers a Master dealing with new technologies is another alternative. We seek to give more attention to students'

IOP Conf. Series: Materials Science and Engineering **291** (2017) 012007 doi:10.1088/1757-899X/291/1/012007

comments gathered from our surveys, among which: creating a digital library at the universities, enhancing design studios with computers, ensuring training sessions on appropriate software and integrating them as part of the curriculum, providing some digital technologies, (as we observed at the ETH Zurich during our internship).
Our aim is to improve students' academic level in accordance with these computer tools issue, not to train technician mastering computer tools. We are rather more involved in the content of the programs and how we can fit the specific courses in order to reach the fixed objectives. The target is to make students able to better use the digital and give suitable attention to their projects rather than focusing only on those tools.
University education undergoes technological improvements. Hence, it needs, not only, to up to date its educational system but has to predict the possible changes in order to evaluate their concrete impact on design, creativity and the architecture in this country. A full understanding and better use of computer tools will certainly help future generations of architects to deal harmoniously with new technologies.

ACKNOWLEDGEMENTS
We greatly thank all students and teachers from the different institutions (Constantine, Biskra, Oran and Algiers) for their effective collaboration.

References

[1] Porada S. 2001-2002. Chantier de l'imaginaire : l'innovation architecturale et l'ordinateur. EAV Enseignement, Architecture, Ville. (7) : 10-13.

[2] Saighi O. and Zerouala M.S. 2012. Information technology utilization in architectural engineering: a field study at the department of architecture and urbanism: Constantine. In: CAAD | INNOVATION | PRACTICE: The 6th International Conference Proceedings of the Arab Society for Computer Aided Architectural Design. pp. 155-156; 273-284.

[3] Lynn G. 2013. Archéologie du numérique: Peter Eisenman, Frank Gehry, Chuck Hoberman, Shoei Yoh. Centre canadien d'architecture. Montréal. p. 33.

[4] Mascaro F. 2004. L'esquisse numérique. Architecture intérieure CREE. (314) : 130-131.

[5] Picon A. 2008. Le projet au risque du numérique. Le Visiteur. (12) : 92-101.

[6] Vivier T. 1997. La souris ou la main. amc. (77) : 89-90.

[7] Madrazo L. 1993. From sketches to computer images: A strategy for the application of computers in architectural design. In : IMARA'93 Colloque (Image animée et representation architecturale). Monte-Carlo. pp. 152-170.

[8] Razavi A. 2008. Digital design. Architecture intérieure CREE. (339) : 54-63.

[9] Lamuniere I. and Marchand B. 2010. Architecture, pure architecture à l'EPFL. L'Architecture d'Aujourd'hui. (377) : 187-196.

Project management practice and its effects on project success in Malaysian construction industry

N A Haron[1], P Devi[1], S Hassim[1], A H Alias[1], M M Tahir[1], and A N Harun[2]

[1]Department of Civil Engineering, Faculty of Engineering, University Putra Malaysia 43400, Serdang, Selangor, Malaysia.

[2]Department of Management of Technology, Malaysia-Japan International Institute of Technology, Universiti Teknologi Malaysia, Jalan Sultan Yahya Petra, 54100 Kuala Lumpur, Malaysia

(nuzul@upm.edu.my, mtahir1129@gmail.com, aidihizami@upm.edu.my)

Abstract: The rapid economic development has increased the demand for construction of infrastructure and facilities globally. Sustainable development and globalization are the new 'Zeitgeist' of the 21st century. In order to implement these projects successfully and to meet the functional aim of the projects within their lifetime, an efficient project management practice is needed. The aim of this study is to identify the critical success factors (CSFs) and the extent of use of project management practice which affects project success, especially during the implementation stage. Data were obtained from self-administered questionnaires with 232 respondents. A mixed method of data collection was adopted using semi-structured interview and questionnaire approach. The result of the analysis of data obtained showed that new and emerging criteria such as customer satisfaction, competency of the project team, and performance of subcontractors/suppliers are becoming measures of success in addition to the classic iron triangle's view of time, cost and quality. An insight on the extent of use of different project management practice in the industry was also achieved from the study.

Keywords: CSFs; Project Management practices; Project success.

1. Introduction

The construction industry is vital since rapid economic development has increased the demand for construction of infrastructure and facilities around the globe. The construction industry also provides the basic living conditions for the sustainability and development of human life on the earth. To cope with an ever-increasing population, pressure on land, and growing economic activity, construction projects are in increasing demand and activities are booming in many countries [26]. More also, projects and initiatives are implemented to ensure sustainable growth of nation economy and to create extensive linkages within the economy. To implement these projects successfully and to meet the functional aim of the projects within their service time, an efficient PM practice needs to be adopted from the planning stage to end.

Frequently, those that are involved in the project handling, fail to take a proactive approach to overcoming the uncertainties [18, 19]. As a result of this, project delays and budget overruns are usually encountered due to an overlook of potential risk. Insufficient information and ineffective management of project not only caused project cost overrun, completion delays but also termination before completion and negatively impact the project team's reputation. To improve the chance of success and reduce the potential failures, the success criteria, and uncertain factors should be carefully identified, assessed and monitored [14].

The Malaysian construction industry plays a vital role in the country's economy, yet it has been plagued with bad publicity of cost overruns, uncontrolled and unrealistic schedules, accidents, poor workmanship, conflict among project team members, abandoned and unfinished private and public construction projects [21]. It is now common to see structures collapsing, roads cracking, bridges toppling and what could be next, show some down pit situations for the construction industry in Malaysia as has left a bad impression on the minds of the public. There is a need and urgency to prevent the failure of projects especially due to poor project management practice in the industry.

Content from this work may be used under the terms of the Creative Commons Attribution 3.0 licence. Any further distribution of this work must maintain attribution to the author(s) and the title of the work, journal citation and DOI.
Published under licence by IOP Publishing Ltd

Nowadays, projects are far more complicated than ever before due to large capital investments, embrace several disciplines, widely dispersed project participants, tighter schedules, stringent quality standards, escalating cost, environment shocks, increasing stakeholders' power and advancement in ICT [2]. Project success may be judged on the basis of how well the resultant product or service supports organizational governance. It is important for the project manager to be knowledgeable about corporate/organizational governance policies and procedures pertaining to the subject matter of the product or service. To ensure the success of the project, the project manager must have the requisite knowledge of project management.

The 12 CSFs commonly related to the implementation of project success identified from a study on 136 European projects that were executed between 1994 and 2004 are categorized into 3 major areas: project management success, individual project success, and corporate success [6]. Table 1 shows some critical factors for successful project accomplishment from the perspective of authors in their studies.

Table 1: Identified Critical Success Factors (CSFs)

Critical Success Factor	Authors
Clear project management objectives	Al-Tmeemy, et al. (2011); Zhang, et al. (2013); Fortune & White (2006); Joyce et al. (2011); Karen, et al. (2010); Khang, et al. (2008); Ofori, (2013).
Top management support	Fortune & White (2006); Hyvari, (2006); Joyce et al. (2011); Khang, et al. (2008); Ofori, (2013); Turner & Muller (2005); Verburg et al. (2013).
Information/ effective communication	Al-Tmeemy et al. (2011); Hyvari, (2006); Joyce et al. (2011); Ofori, (2013); Verburg et al. (2013).
Client involvement	Fortune & White (2006); Hyvari, (2006); Joyce et al. (2011); Khang, et al. (2008); Turner & Muller (2005).
Competency of project team	Zhang, et al. (2013); Fortune & White (2006); Joyce et al. (2011); Ofori, (2013); Verburg et al. (2013).
Authority of the project manager/leader	Zhang, et al. (2013); Fortune & White (2006).
Realistic cost and time estimates	Abu Hassan et al. (2009); Akewushola et al. (2012); Al-Tmeemy et al. (2011); Fortune & White (2006).
Adequate project control	Fortune & White (2006).
Problem solving abilities	Zhang, et al. (2013); Hyvari, (2006).
Project performance and quality	Akewushola et al. (2012); Al-Tmeemy et al. (2011); Karen et al. (2010).
Adequate resources	Abu Hassan et al. (2009); Al-Tmeemy et al. (2011); Hyvari, (2006); Joyce et al. (2011); Khang, et al. (2008).
Effective planning	Fortune & White (2006); Joyce et al. (2011); Khang, et al. (2008); Turner & Muller (2005).
Monitor performance and feedback	Fortune & White (2006); Hyvari, (2006); Joyce et al. (2011); Turner & Muller (2005).
Project mission/common goals	Al-Tmeemy et al. (2011); Hyvari, (2006); Khang, et al. (2008); Turner & Muller (2005); Verburg et al. (2013).
Adequacy of contingency plan (Risk management)	Joyce et al. (2011); Ofori, (2013).
Customer satisfaction	Abu-Hassan et al. (2009); Akewushola et al. (2012); Al-Tmeemy et al. (2011); Zhang, et al. (2013); Ofori, (2013).
Well-laid out specification	Joyce et al. (2011); Ofori, (2013).
Effective selection/use of technology	Joyce et al. (2011); Verburg et al. (2013).

Developing a more comprehensive framework is essential to improve the success rate of projects since more organizations are expected to manage multiple projects in order to achieve competitive advantages [22]. A very comprehensive project management framework should consider other elements to include cultural, structural, practical, and personnel [3]. Due to the myriad challenges faced by this construction industry. The aim of the study is to identify and assess the critical success factors (CSFs) and the current practice of project management that affects project success at the implementation stage.

2. Methodology

A mixed method of enquiry was adopted for a better understanding of the behavior of Malaysia construction industry in managing a project. Data were collected using a quantitative approach, where a systematic empirical investigation of observable phenomena via statistical, mathematical or computational techniques [27] involved. A questionnaire comprising 16 closed ended questions were developed with a mixture of multiple choices, five-point Likert scale options [14].

After finalizing the questionnaire, a pilot test was undertaken with 30 targeted respondents to ensure that respondents would understand the questions and identify possible problems with the completion of the questionnaire. The widely acceptable response rate in the construction industry for electronic questionnaire is 20 – 30% [4, 7]. The respondents were drawn from the G1 to G7 construction companies in Klang Valley in Selangor state in Malaysia. A total of 378 questionnaires were administered, and only 232 were returned, which represented 61.4% of the expected population. The questionnaire comprised of four sections.

3. Results

In this research, Statistical Package of Social Sciences-SPSS 20 software was utilized for both quantitative and qualitative methods, with cross-tabulations to clarify the relationships between the variables [16]. Basic descriptive analysis was conducted to find the standard deviation of the variable. The aim is to find the most relevant variables in the specified category. Furthermore, the results use as a baseline for comparing independent variables.

The reliability of multiple Likert scale questions was measured using Cronbach's alpha. The alpha value obtained was 0.928 which shows a high level of internal consistency for the scale with this specific sample. Goodman and Kruskal's λ was also run to determine whether the three (3) related variables could better predict by class of contractors and the findings shown in Table 2. The Goodman and Kruskal's λ were 0.110 for the maturity level of PM, 0.057 for organization support and 0.175 for project success. These indicate that there are some proportional reduction in errors in predicting the followed dependent variables when taking into account the class of contractor as an independent variable. However, there was no statistically significant decrease in the proportion of errors due to the tested dependent variables as predictors of the class of contractor.

This indicates that any scale of construction firms should choose appropriate PM practice based on project needs, and its success relies on how well the practices being utilized during the implementation stage. Based on Table 2, the maturity level of project management identified is at "defined level" and the project success rate is average to more successful.

Table 2: Level of Project Management Practice, Organization Support and Rate of Project Success

Variable	Nominal by Nominal: Lambda			
Dependent Variable	Value	Asymp. Std. Error	Approx. T^b	Approx. Sig.
Maturity level of PM	0.110	0.074	1.408	0.159
Organization support	0.057	0.040	1.406	0.160
Rate of project success	0.175	0.055	2.945	0.003

Table 3: Descriptive Statistics for Contractors' Critical Success Factors (CSFs)

Critical Success Factors (CSFs)	N of Valid cases	Mean	Std. Dev.	Symmetric Measures		Rank
				Value	Approx. Sig.	
Financial attributes (Turnover	223	4.36	.754	-0.08	0.366	3

history, credit history, etc.)						
Effective planning and controlling	222	4.38	.683	0.022	0.808	2
Realistic cost and time estimation	223	4.32	.626	-0.051	0.571	4
Competency of project team	223	4.30	.702	-0.037	0.672	5
Adequacy of resources (Labour, plant, etc.)	221	4.26	.670	-0.09	0.322	7
Contractual awareness	221	4.16	.681	0.05	0.579	10
Past projects' experience and performance records	222	3.99	.751	0.083	0.323	13
Organization attributes (Image, length of time in business, etc.)	222	3.76	.781	0.09	0.274	15
Top management support	222	4.21	.717	0.131	0.119	8
Performance of subcontractors/suppliers	222	4.29	.717	0.205	0.016	6
Total quality management	220	4.17	.724	-0.016	0.851	9
Environmental, health and safety attributes	223	4.09	.775	-0.161	0.054	11
Customer satisfaction	223	4.44	.669	-0.106	0.23	1
Managing new technologies	223	3.86	.694	-0.003	0.968	14
Mutual learning/knowledge sharing	223	4.03	.677	-0.051	0.581	12

A total of fifteen (15) CSFs for project implementation were identified for this study. The targeted respondents were asked to indicate the level of importance for each factor as shown in Table 3. The response rate for each CSFs (220 – 223 responses or 94.83 – 96.12%). Goodman and Kruskal's was run to determine the association between a list of critical factors and class of contractor. The results in Table 2 showed a weak association between the critical success factors and class of contractor, which were not statistically significant (P > 0.0005). This indicates the representative respondents from the various class of contractor do not affect in choosing CSFs that can contribute to the success of the project. Table 3 shows the mean and standard deviation values for each CSFs. The twelve (12) CSFs had a mean average between 4.03 to 4.44, this indicates that the mean responses to this questions were Important, with the exception of C7, C8, and C14, which had average mean values of 3.99, 3.76 and 3.86which are also moderately important.

In order to assess the extent of use of the project management standards, methods, methodologies, tools, and techniques that are widely used, questions were asked in order to measure the extent of use for twenty-three (23) research variables which grouped into four (4). This is shown in Table 4 below.

Table 4. Descriptive Statistics for Extent of Use of PM Practices

Item	No of Valid Cases	Mean	Std. Dev.	Symmetric Measures	
				Value	Approx. Sig.
Project Management Institute (PMI)	215	2.91	1.151	0.264	0.000
International Project Management Association Competence Baselines (IPMA)	214	2.79	1.219	0.327	0.000
Projects in Controlled Environments (PRINCE/ PRINCE2)	215	2.82	1.201	0.302	0.000
ISO10006.2003	215	3.02	1.298	0.338	0.000
In-house project management methods	220	3.34	1.079	0.289	0.000
Other project management methods	160	2.71	1.355	0.224	0.006
Cost benefit analysis (CBA)	218	3.61	.909	0.128	0.114

Decision analysis (DA)	216	3.42	1.004	0.171	0.039
Sensitivity analysis (SA)	217	3.24	1.046	0.25	0.001
Expressed preferences	218	3.27	1.043	0.178	0.023
In house decision making techniques	219	3.48	1.085	-0.025	0.771
Other decision making techniques	157	2.82	1.346	-0.003	0.972
Critical Path Method (CPM)	217	3.37	1.198	0.296	0.000
Work Breakdown Structure (WBS)	214	3.33	1.159	0.253	0.001
Gantt bar chart	216	3.44	1.148	-0.397	0.000
Project management software	215	3.21	1.202	0.323	0.000
In-house project management tools	217	3.32	1.164	0.212	0.008
Other project management tools	156	2.58	1.344	0.045	0.580
Life-Cycle Cost Analysis (LCCA)	215	3.09	1.144	.333	0.000
Event tree analysis	214	3.02	1.150	.227	0.002
Probability Analysis (PA)	214	3.01	1.190	.304	0.000
In-house risk assessment tools	217	3.19	1.156	.156	0.047
Other risk assessment tools	158	2.59	1.302	.046	0.581

The association between the tested PM practices was ranging from weak to moderate and not all the variables were statistically significant. These indicate that the usage of some variables depends on size and complexity of projects. Choosing appropriate variables, guide the organization to achieve their goals easily. Meanwhile, the extent of usage of each variable ranges from "rarely" to "occasionally", which shows that the industry less utilized PM practice for project implementation due to lack of knowledge and exposure based on real successful case studies. The best use of PM practice associated with the size and complexity of projects and choosing appropriate practices, guide the construction firms to achieve their goals easily.

4. Discussion of Result

There is a more considerable dispute about PM practice and its contribution towards project success. Any agreement has not been reached, even the topic has discussed for an extended period. The study identified the top five factors that influence the project success to include customer satisfaction, effective planning and controlling, financial attributes, realistic cost and time estimation, and competency of the project team. The concept of CSFs presents a smarter way to identify certain factors which could be possible to make the project successful.

Applying PM practice has become important issues in many developed countries due to its successful application in various industries and its proven effectiveness and flexibility in attaining project goals and objectives. Due to its nature with high risk and consuming many resources, construction industry requires better application and utilization of efficient and effective PM practice. Studying the use of PM standards, methods, methodologies, tools and techniques that widely applied in this industry, serves as eye openers to the contractors and another decision maker to better plan their effort toward the efficient application of PM practice. If properly utilized, PM practice would result in concrete benefits in all aspects of project implementation.

The result from the study showed that the limited use of PM practice, an ad hoc approaches were preferred due to high cost, lack of expertise in PM and difficulty in real world modeling. CBA, Gantt Bar Chart, and CPM are some of the most widely applied PM tools and techniques by the respondents because of their simple and user-friendly nature. To tackle these crucial obstacles, the respondents stated that adequate training should be given to the employee in the art of PM and on real successful case studies. However, contractors are main players related to construction projects and their success rely on the actual work of the project accomplished. The findings from this study provide a clear understanding of contractors' role and potential construction project success.

5. Conclusion and Recommendation

The following are the conclusions and recommendation for proper application of PM practice in Malaysian construction industry in general:

- PM tools and techniques should be applied gradually especially in small-scale firms. The drastic use should be avoided so as not to lead to a disruptive change in their business.
- Adequate PM training or courses should be conducted in higher education institutions, to increase the knowledge of practitioners about different PM tools and techniques available for appropriate choice at the early stage.
- The government should step up its commitment in applying PM practice in all governments' projects, so as not to become inefficient and unproductive. A special allocation should be prepared to enhance the efficient and effective application of PM tools and techniques among contractors.
- The undertaking of construction projects in metropolitan regions is a risky competitive and dynamic due to their surrounding environments, which are complicated in term of transportation, the number of direct and indirect stakeholders involvement and the handling of existing facilities [14].
- The lack of documentation on project management in construction industry reduced the data available that could have incorporated into the study.
- The study only focused on the extent of use of the most common PM practice that used for project implementation in this country.
- The construction activity has a greater impact on the environment than other industries, an urgent need to apply sustainable development principles to construction industry practices [26]. Meanwhile, further study can be carried out in the same context by including other variables for project implementation or applying for different phases of the industry that might have a significant effect on the project success.

References

[1] Abu Hassan et al. (2009). Project Management Success Factors for Sustainable Housing: A Framework. Retrieved.
[2] Adeyemi, I. (2013). Effects of Project Management on the Performance of a Construction Firm in Nigeria. American International Journal of Contemporary Research, 3(6), 54-58.
[3] Akewushola et al. (2012). Effect of Project Management on Project Success. Australian Journal of Business and Management Research, 2(3), 1–11.
[4] Akintoye, A. (2000). Analysis of factors influencing project cost estimating practice. Construction Management and Economics, 18 (1), pp 77-89.
[5] Al-Tmeemy, Abdul-Rahman, & Harun, (2011). Future criteria for success of building projects in Malaysia. International Journal of Project Management, 29(3), 337–348.
[6] Bash, Eleanorv (2015). The project success rate and standard project management methodology in Malaysia. PhD Proposal.
[7] Dulami, M.F., Ling, F.Y.Y. and Bajracharya (2003). Organisational motivation and inter-organisational interaction in construction innovation in Singapore. Construction Management and Economics, 21 (3), pp.307-318.
[8] F. Zhang, Zuo, & Zillante (2013). Identification and evaluation of the key social competencies for Chinese construction project managers. International Journal of Project Management 31, 748– 759.
[9] Fortune & White (2006). Framing of project critical success factors by a systems model. International Journal of Project Management, 24(1), 53–65.
[10] Hyvari, I. (2006). Success of Projects in Different Organizational Conditions. Project Management Journal, 37(4), 31–41.

[11] Joyce et al. (2011). Looking again at current practice in project management, International Journal of ManagingProjects in Business, 4 (4), 553 – 572.
[12] Karen et al. (2010). Do project managers practice what they preach, and does it matter to project success?, International Journal of Project Management, 28 (7) 650–662.
[13] Khang, D.B., Moe, T.L., 2008. Success criteria and factors for international development projects: a life-cycle-based framework. Project Management Journal 39 (1), 72–84.
[14] Kuo, Y.-C., & Lu, S.-T. (2013). Using fuzzy multiple criteria decision making approach to enhance risk assessment for metropolitan construction projects. International Journal of Project Management, 31(4), 602–614.
[15] Nunnally JC. Psychometric theory. New York: McGraw-Hill; 1978
[16] Ofori, D. F. (2013). Project Management Practices and Critical Success Factors–A Developing Country Perspective. International Journal of Business and Management, 8(21), 14–32.
[17] Pallant, J. (2001). SPSS Survival Manual. A step by step guide to data analysis using SPSS for windows (Version 10-11). Buckingham: Open University Press.
[18] Project Management Institute [PMI]. (2006). A Guide to the Project Management Body of Knowledge. USA: Project Management Institute (PMI).
[19] Project Management Institute [PMI]. (2008). A Guide to the Project Management Body of Knowledge. USA: Project Management Institute (PMI).
[20] Takim, R. (1999). Analysis of Effectiveness Measures of Construction Project Success in Malaysia, 4(7), 74–91.
[21] Ting, S., Khoo, H., & Wong, S. (2009). Project Management Development in Malaysia: A Case Study. Department of Civil Engineering, University Malaysia.
[22] Too, E. G., & Weaver, P. (2014). The management of project management: A conceptual framework for project governance. International Journal of Project Management, 32(8), 1382–1394.
[23] Turner, J. R., & Müller, R. (2005). The Project Manager's Leadership Style as a Success Factor on Projects: A Literature Review. Project Management Journal, 36(2), 49–61.
[24] Uma Sekaran, Roger Bougie, (2013), Research Methods for business: A skill-building approach, 6th edition, Wiley.
[25] Verburg, R. M., Bosch-Sijtsema, P., & Vartiainen, M. (2013). Getting it done: Critical success factors for project managers in virtual work settings. International Journal of Project Management, 31(1), 68–79.
[26] Zhang, X., Wu, Y., Shen, L., & Skitmore, M. (2014). A prototype system dynamic model for assessing the sustainability of construction projects. International Journal of Project Management, 32(1), 66–76.
[27] Given, L. M. (Ed.). (2008). The Sage encyclopedia of qualitative research methods. Sage Publications.

Applications of building information model (BIM) in Malaysian construction industry

M M Tahir[1], N A Haron[1], A H Alias[1], A T Al-Jumaa[1], I B Muhammad[2] and A N Harun[3].

[1]Department of Civil Engineering, Faculty of Engineering, Universiti Putra Malaysia, Malaysia

[2]Department of Architecture, School of Environmental Technology, Federal University of Technology, Minna, Nigeria

[3]Department of Management of Technology, Malaysia-Japan International Institute of Technology, Universiti Teknologi Malaysia, Jalan Sultan Yahya Petra, 54100 Kuala Lumpur, Malaysia

(mtahir1129@gmail.com, nuzul@upm.edu.my, aidihizami@upm.edu.my, azher_taha1978@ymail.com, mibdoko@gmail.com)

ABSTRACT: Since the introduction of BIM in Malaysia in 2009, the technology adoption rate is slow when compared to other countries of the world. Most of the construction companies in Malaysia have an insight on the BIM concept but are yet to implement it in the management of their construction projects. By the year 2020, the Malaysian government will make BIM mandatory, this makes it important to carry out research on the possible applications of the technology. A qualitative method of enquiry was used for this study in Klang Valley using semi-structured interview. The responses received were analysed using Principal component analysis (PCA). The result of the analysis showed that "quantity take-off and estimation", "clash detection and coordination", "integration and collaboration of stakeholders", and "design and visualisation" as the main applications of BIM in Malaysia presently. The implication of this findings is that the Malaysian construction industry productivity is likely to increase to meet the demand of the population through the implementations of BIM. More also, BIM technology is regarded as the future of construction industry, which makes it very important for the industry.

Keywords: Applications of BIM, BIM, Malaysian construction industry.

1. INTRODUCTION

Building Information Model (BIM) in construction design is regarded as an alternative approach for design which makes it easy to present digital designs which contains all the necessary information about the proposed project before it is constructed [1]. BIM is an innovative design tool which has changed a lot in the construction industry such as project procurement, execution, and facility management [2]. This shows that BIM can be said to be the future of construction industry.According to Bryde, Broquetas [3], BIM is a suitable tool for management of construction projects, as such, project managers should take advantage of it in managing their projects. More also, Enegbuma, Ologbo [4] stated that BIM reduces uncertainties which result in successful completion of projects. This has so far proven to be the most important benefit of BIM application. It is important to note that BIM is applicable at all stages of a project (Initiation to operation and maintenance) [5].

A widely cited definition of BIM is provided by the US National Institute of Building Sciences (2007) which defined it as "a digital representation of physical and functional characteristics of a facility, and a shared knowledge resource for information about a facility forming a reliable basis for decisions during its life-cycle; defined as existing from earliest conception to demolition". For the purpose of this research, the definition credited to the National BIM Standard (NBIMS) Project Committee of the Building SMART alliance (2010) will be adopted, which defines it as, "a digital representation of physical and functional characteristics of a facility. As such it serves as a shared knowledge resource for information about a facility forming a reliable basis for decisions during its life-cycle from inception onward. The BIM is a shared digital representation founded on open standards for interoperability" [6].

Content from this work may be used under the terms of the Creative Commons Attribution 3.0 licence. Any further distribution of this work must maintain attribution to the author(s) and the title of the work, journal citation and DOI.
Published under licence by IOP Publishing Ltd

BIM is basically a 3D digital representation of a facility. The model could be used in expressing the entire facility life-cycle. The quantity of material and its properties can be easily obtained, and the scope of work required can easily be defined and isolated from the model. Contract documents, drawings, procurement details, specifications, and other construction documents can easily be interrelated using the model [7, 8]. The model which is data rich, object-oriented, intelligent, and a digital representation of the facility in which drawings and appropriate information for various stakeholders can be extracted for project delivery and decision-making [9]. It integrates architectural, structural, Mechanical, Electrical and Plumbing (MEP) models [10].

In practice, the information contained in the model do not always show the real life situation which limits its application in construction projects. The model is disconnected from the physical project which reduces its usefulness throughout the facility life-cycle in supporting decisions [11]. This is as a result of lack of framework to guide practitioners and researchers to come up with solutions to link BIM and physical facilities. The implementation of this model by most AEC companies is as a result of it benefit for long-term and productivity gain when compared to conventional practices [12]. The potentials of the application of Building Information Model (BIM) in the management of construction projects is similar to the PMBOK knowledge areas, this makes BIM an important tool for project managers [13]. BIM as a promising technology facilitates project management, the possibility of integrating building models and products makes BIM have a high potential for management of projects life-cycle [14]. BIM can be used at all the stages of a project in it life-cycle, it is used in understanding the project needs by the owner, it is also used for analysis, design and development of the project by the design team.

However, since the introduction of BIM in Malaysia in 2009, the technology adoption rate is slow when compared to other countries of the world. Most of the construction companies in Malaysia have an insight of the BIM concept, but due to the absence of guidance, government support, and well-trained personnel have been the major cause for the slow rate of adoption of the technology [9]. This is considered as a major setback for the Malaysian construction in adopting BIM when compared to US which crossed the gap within 5 years of the introduction of BIM to their industry [15]. In November 2016, the chief executive officer of CIDB, Datuk Ahmad Asri Abdul Hamid, made a press statement saying "property developers will be mandated to use BIM by 2020 for their construction works" [16]. This is a good development for the industry, but however, BIM as an alternative technology in Malaysia needs to be studied to know it applications and possible areas of improvement in the industry. More also, the BIM maturity level appreciation in Malaysia with regard to increasing demand for efficiency and competitive advantages is been hindered by several factors [17] which include low research on applications of BIM in the management of construction projects.

2. METHODOLOGY

A qualitative method of research was used for this study. Qualitative research makes use of subjective data which are produced from the respondent's minds (Creswell, 2013). The purpose of the qualitative approach was to reaffirm findings of the applications of BIM from the literature and to uncover other applications in Malaysia. The study was conducted using semi-structured interview. The targeted respondents for the interview were architects, engineers, project managers, and BIM coordinators who are experts and have knowledge and understanding of the applications and benefits of BIM. Expert sampling method was used in selecting respondents from the population in Malaysia (Klang Valley). The data obtained from the interview were reported and analysed using Descriptive analysis, Cronbach's alpha test, Exploratory Factor analysis, and Content analysis by carefully constructing data categories.

3. RESULTS

Data were obtained through the use of semi-structured interview with 18 respondents who were experts and have extensive knowledge about BIM. The respondents were architects, engineers, BIM coordinators, and project managers. Prior to the interview, a pilot study was conducted with 3 respondents, in order to ensure that respondents understand the questions asked. Data from the pilot study was used to restructure the interview questions for better understanding of the respondents.

3.1 Demographic Characteristics of the Respondents

The respondents for the semi-structured interview were from different backgrounds in terms of their job positions, industry working experience, BIM working experience, and their level of education. Descriptive statistics was used in examining the different distributions of variables from their backgrounds as shown in Table 1 below.

Table 1: Frequency distribution of Respondent's demographic characteristics

Variable	Item	Frequency	Percentage
Job description	Architect	6	33.3
	BIM coordinator	3	16.7
	Engineer	4	22.2
	Project manager	5	27.8
Industry working experience	Less than 5years	4	22.2
	5-10 years	8	44.4
	More than 10 years	6	33.3
Working experience with BIM	Between 1& 2 years	4	22.2
	3-5 years	10	55.6
	More than 5 years	4	22.2
Highest level of qualification	BSc	3	16.7
	MSc	13	72.2
	PhD	2	11.1

Table 1 above shows the summary of the background of the respondents. The result shows the highest number of respondents to be architects with 33.3% followed by project managers with 27.8%. The other respondents were Engineers and BIM coordinators having 22.2% and 16.7% respectively.

However, the result of the respondent's industry working experience shows that 44.4% of the respondents have been working in the industry between 5-10 years. While 33.3% of the respondents have spent more than 10 years in the industry, and only 22.2% of the respondent have spent less than 5 years in the industry. More also, 55.6% of the respondents had 3 to 5 years working experience with BIM in their various projects, while the respondents with more than 5 years working experience with BIM and those that had between 1 and 2 years working experience with BIM had 22.2% each. The level of education of the respondents showed that they are qualified to participate as respondents in the survey as 72.2% of them had Master's degree, 16.7% had Bachelor degree, and 11.1% of the respondents had PhD degree in their various professions. All of these characteristics of the respondents shows that they are all qualified to respond to the semi-structured interview questions as experts.

3.2 Reliability Analysis

The reliability test was conducted on the questions using Cronbach's alpha test to measure the questions reliability. The reliability analysis results are shown in Table 2.

Table 2 Reliability analysis on applications of BIM

Application of BIM	Scale Mean if Item Deleted	Corrected Item-Total Correlation	Cronbach's Alpha if Item Deleted	Cronbach's Alpha Coefficient
Design and visualisation	43.89	0.017	0.803	
Scheduling	44.89	0.723	0.743	
Quantity take-off and estimation	44.11	0.477	0.771	
Integration and collaboration of stakeholders	44.50	0.387	0.780	
Clash detection and coordination	44.22	0.341	0.783	
Constructability analysis	44.56	0.608	0.760	0.790
Quality and risk control	45.06	0.614	0.754	
Facility management	44.39	0.702	0.752	
Communication	45.06	0.623	0.752	
Human and material resource control	45.28	0.677	0.745	
Reduction of rework during construction	44.50	0.827	0.746	
Contract processes	44.89	-0.445	0.864	

The scale reliability and internal consistency of data from this phase of the study are shown in table 2. The scale reliability and consistency for the 12 items tested which is the alpha coefficient is 0.790 as shown in table 2. This implies that the internal consistency of the items is relatively high. It is important to note that the minimum acceptable reliability coefficient is 0.70.

3.3 Data Analysis:

The information obtained from the semi-structured interview were categorised and tabulated using content analysis. Descriptive analysis was used to describe the nature of the data before been analysed using SPSS 23. The response from the experts was interpreted using scaling method for easy coding. The scale which ranges from strongly agree to strongly disagree.

The responses were further analysed using exploratory factor analysis to show that the items a drawn from the concept being discussed as shown in Table 3. The Kaiser-Meyer-Olkin Measure of sampling adequacy was not performed on the items due to the number of respondents that participated in the semi-structured interview.

Table 3: Communalities of the items

BIM application	Initial	Extraction
Design and visualisation	1.000	0.849
Scheduling	1.000	0.800
Quantity take-off and estimation	1.000	0.867
Integration and collaboration of stakeholders	1.000	0.886
Clash detection and coordination	1.000	0.796
Constructability analysis	1.000	0.741
Quality and risk control	1.000	0.736
Facility management	1.000	0.873
Communication	1.000	0.910
Human and material resource control	1.000	0.705
Reduction of rework during construction	1.000	0.841
Contract processes	1.000	0.832

The communalities table represent the amount of variance accounted for by the factor solution for each variable. These shows whether if the variables meet an acceptable level. The minimum acceptable level for these variables is set at 0.50 variance which must be taking into account. In this phase of the study, the communalities were all above (0.50) as shown in Table 3 above.

Table 4: Total Variance Explained

	Initial Eigenvalues			Extraction Sums of Squared Loadings			Rotation Sums of Squared Loadings		
Comp.	Total	% of Variance	Cum. %	Total	% of Variance	Cum. %	Total	% of Variance	Cumulative %
1	5.492	45.766	45.766	5.492	45.766	45.766	3.051	25.429	25.429
2	1.826	15.217	60.983	1.826	15.217	60.983	2.862	23.852	49.280
3	1.364	11.366	72.349	1.364	11.366	72.349	2.684	22.369	71.649
4	1.154	9.615	81.965	1.154	9.615	81.965	1.238	10.316	81.965
5	.680	5.664	87.628						
6	.526	4.384	92.012						
7	.421	3.510	95.522						
8	.220	1.830	97.352						
9	.140	1.168	98.520						
10	.103	.854	99.375						
11	.064	.529	99.904						
12	.012	.096	100.00						

Extraction Method: Principal Component Analysis.

The communalities of all the variables were greater than 0.5 which allows them all (12 items) to be used for the factor analysis. Four components with eigenvalues greater than (1) were found. The eigenvalues and total variance explained by the four components is shown in table 4 above. The Eigenvalues after Varimax rotation showed that the first factor explained (45.77 %) of the variance, the second factor (15.22%) of the variance, the third factor (11.37%) of the variance, while the fourth factor explained (9.62%) of the variance. These four components explain (81.97%) factor structure of the total variance among the items.

Table 5: Rotated Component Matrix[a]

	Component			
	1	2	3	4
Design and visualisation	-	-	-	.914
Scheduling	.743	.450	-	-
Quantity take-off and estimation	.917	-	-	-
Integration and collaboration of stakeholders	-	-	.913	-
Clash detection and coordination	-	.735	-	.454
Constructability analysis	.835	-	-	-
Quality and risk control	-	.655	.400	-
Facility management	.477	.694	-	-
Communication	-	-	.894	-
Human and material resource control	.537	-	.619	-
Reduction of rework during construction	.494	.537	.555	-
Contract processes	-	-	-	-

Extraction Method: Principal Component Analysis. Rotation Method: Varimax with Kaiser Normalization.

The "Varimax rotational matrix[a]" in Table 5 above shows the correlation of each of the items with the four components with eigenvalues greater than (1) that were selected. The correlation relationship ranges from (+1) to (-1) indicating the strength of the relationship. These components can be named after the items that show very high correlation with them when compared to the other items. The first component has its highest correlation with the item "Quantity take-off and estimation". The second component has its highest correlation with the item "Clash detection and coordination". The third component has its highest correlation with the item "Integration and collaboration of stakeholders". While the fourth component has its highest correlation with the item "Design and visualisation".

4. DISCUSSION

Twelve (12) items were identified from the literature to be the main applications of BIM in the management of construction projects. These items are design and visualisation, scheduling, quantity take-off and estimation, integration and collaboration of stakeholders, clash detection and coordination, constructability analysis, quality and risk control, communication tool, facility management, human and material resource control, contract process, and reduction of rework during construction.

These items were validated through the use of semi-structured interview which was further analysed using statistical methods (descriptive analysis, factor analysis). The analysis of the items using factor analysis showed that four items were the most important applications of BIM as they were selected as components. These items are quantity take-off and estimation, clash detection and coordination, integration and collaboration of stakeholders, and lastly design and visualisation. These findings can be referenced to researches conducted previously by different researchers.

The result of the study shows that quantity take-off and estimation is one of the main application of BIM in the management of construction projects. This concurs with the findings of [18-20], Lahdou and Zetterman [21], and Sabol, (2008) that stated that cost estimation and quantity take-off are the key applications of BIM in project management which are continuously developed.

The result of the research also showed that BIM is been applied in the management of construction projects to detect clash in designs and coordination of work. Bryde, Broquetas [3] stated that one of the most reported applications of BIM was is to improved coordination throughout the project life-cycle. More also, Farnsworth, Beveridge [22] stated that the applications of BIM to detect clashes in designs and coordinate activities reduces the time spent on construction and at the same time increases profitability and marketing.

The integration and collaboration of stakeholders was also identified as one of the main applications of BIM in the management of construction projects. The use BIM reduces waste and also optimises efficiency throughout the project life-cycle as it supports integrated project delivery through collaborative process [23].

The application of BIM for design and visualisation in the management of construction project was also identified in the research. BIM is a great visualisation tool as it provides a 3D virtual representation of the facility [24]. Walkthrough, rendering, and sequence of the model (BIM) can be provided by the project manager during the project bidding for ease in communicating with the interested contractors. An outlook of the project when completed is provided through visualisation which solves the issues of having to combine the different 2D views of the proposed project to create a 3D view [25].

5. CONCLUSIONS

The adoption of BIM in the management of construction projects is rapidly gaining acceptance globally. However, Malaysia is not left out from these countries. The adoption rate in Malaysia has witnessed a great setback since the introduction of the alternative technology (BIM) in 2009. This is as a result of low research and active government and client participation in the adoption process. This research focused on the main applications of BIM globally to identify the current application in Malaysia. Four items were identified as the main applications of BIM in Malaysia, other uses of the technology are also used but not as much as the identified items. The findings of this research cannot be used to generalise the application of BIM in Malaysia due to the number of respondents used for the study and the limited time spent on the research.

6. REFERENCES

[1] Azhar, S., *Building information modeling (BIM): Trends, benefits, risks, and challenges for the AEC industry.* Leadership and Management in Engineering, 2011. **11**(3): p. 241-252.
[2] Xiao, H. and T. Noble. *BIM's impact on the project manager.* in *Proceedings of the 30th Annual ARCOM Conference*. 2014.
[3] Bryde, D., M. Broquetas, and J.M. Volm, *The project benefits of building information modelling (BIM).* International Journal of Project Management, 2013. 31(7): p. 971-980.
[4] Enegbuma, W., et al. *Preliminary Study Impact of Building Information Modelling Use in Malaysia.* in *IFIP International Conference on Product Lifecycle Management*. 2014. Springer.
[5] Memon, A.H., et al., *BIM in Malaysian construction industry: status, advantages, barriers and strategies to enhance the implementation level.* Research Journal of Applied Sciences, Engineering and Technology, 2014. **8**(5): p. 606-614.
[6] Succar, B., *Building information modelling framework: A research and delivery foundation for industry stakeholders.* Automation in construction, 2009. **18**(3): p. 357-375.
[7] Bazjanac, V., *Virtual building environments (VBE)-applying information modeling to buildings.* 2006.
[8] Khemlani, L., *Top criteria for BIM solutions.* A survey conducted by AECbytes, 2007.
[9] Rogers, J., H.-Y. Chong, and C. Preece, *Adoption of building information modelling technology (BIM) perspectives from Malaysian engineering consulting services firms.* Engineering, Construction and Architectural Management, 2015. **22**(4): p. 424-445.
[10] Smith, D.K. and A. Edgar, *Building information modeling (BIM).* Natonal Institute of Building Sciences, Washington, 2008.
[11] Chen, K., W. Lu, and Y. Peng. *A Preliminary Study on the Framework and Technologies for Bridging BIM and Building.* in *Proceedings of the 19th International Symposium on Advancement of Construction Management and Real Estate*. 2015. Springer.
[12] Eadie, R., et al., *A survey of current status of and perceived changes required for BIM adoption in the UK.* Built Environment Project and Asset Management, 2015. **5**(1): p. 4-21.
[13] Rokooei, S., *Building Information Modeling in Project Management: Necessities, Challenges and Outcomes.* Procedia-Social and Behavioral Sciences, 2015. **210**: p. 87-95.
[14] Gourlis, G. and I. Kovacic, *Building Information Modelling for analysis of energy efficient industrial buildings–A case study.* Renewable and Sustainable Energy Reviews, 2016.
[15] Porwal, A. and K.N. Hewage, *Building Information Modeling (BIM) partnering framework for public construction projects.* Automation in Construction, 2013. **31**: p. 204-214.
[16] Růžku, R.M., *Developers must use BIM system by 2020*, in *BH Online*. 2016: Kuala Lumpur.

[17] binti Ali, K.N. and T.C. Boon, *Building Information Modelling Awareness and Readiness*. 2013.

[18] Azhar, S., M. Hein, and B. Sketo, *Building Information Modeling (BIM): Benefits, Risks and Challenges. McWhorter School of Building Science*. Auburn University. Auburn. Alabama, AL, 2008.

[19] Hergunsel, M.F., *Benefits of building information modeling for construction managers and BIM based scheduling*. 2011, Worcester Polytechnic Institute.

[20] Sabol, L., *Challenges in cost estimating with Building Information Modeling*. IFMA World Workplace, 2008.

[21] Lahdou, R. and D. Zetterman, *BIM for Project Managers How project managers can utilize BIM in construction projects*. 2011.

[22] Farnsworth, C.B., et al., *Application, Advantages, and Methods Associated with Using BIM in Commercial Construction*. International Journal of Construction Education and Research, 2015. **11**(3): p. 218-236.

[23] Glick, S. and A. Guggemos. *IPD and BIM: benefits and opportunities for regulatory agencies*. in *Proc., 45th Associated Schools of Construction National Conference*. 2009.

[24] Zhang, D., *Project Time and Cost Control Using Building Information Modeling*, in *Construction Management and Engineering*. 2012, North Dakota State University of Agriculture and Applied Science ProQuest LLC. p. 46.

[25] Mohandes, S.R., et al., *Comparison of Building Existing Partitions through Building Information Modeling (BIM)*. Jurnal Teknologi, 2015. **75**(1).

Sustainability knowledge using "AKASA" model among architecture students from Klang Valley private universities, Malaysia.

Sivaraman Kuppusamy[1], Mohd Faris Khamidi[1], Lee Xia Sheng[1] and Tamil Salvi Mari[2]
[1]School of the Built Environment, University of Reading Malaysia, Johor, Malaysia
[2]School of Architecture, Building & Design, Taylors's University Lakeside, Selangor, Malaysia
E-mail: s.kuppusamy@reading.edu.my

Abstract. The study intend to investigate sustainability knowledge using "AKASA" model. This model comprises all the literacy level which is the awareness, knowledge, attitude, skills and action. 234 students from 5 selected private universities were surveyed using questionnaires. Students were specifically selected from year 2 and year 3 from private universities in Klang valley, Malaysia. The study intends to investigate the environmental literacy level specifically the knowledge variable. The parametric study was conducted with descriptive analysis and the results shows that the environmental knowledge is at high level compared to other environmental literacy variables among year 2, year 3 and combine year 2 and year 3.

1. Introduction

The current education system is not helping the global challenges towards nature and the current environmental crisis is reflecting the current environmental education crisis. To address the issue, universities worldwide are integrating environmental literacy in the general education particularly in architecture schools, as they are involved with the built environment which instructs the undergraduates to design in correspondence to the site context and community, deal with building materials and create spaces for new ways of living [1].

However, studies on environmental literacy in tertiary education in Malaysian architectural courses, are still scarce. Much of this environmental literacy is being integrated into our subject modules at only surface level, resulting in mediocre outcomes for an eco-literate society and individuals. Research in environmental literacy in architectural courses, reflected on how the environmental studies and thinking are being taken lightly and has not gone into the deeper depths of learning framework and practice [2]. Lembaga Arkitek Malaysia (LAM) established, Council of Accreditation and Architectural Education Malaysia (CAAEM) which regulates and recognizes all matters relating to architectural education has aimed to produce competences that an architecture students need to achieve and recognizes design capabilities, knowledge, and skills to fulfill an architect's responsibilities who can co-ordinate interdisciplinary objectives. This method provides "a synthesis of knowledge, aptitudes, and attitudes" [3]. The study intends to find the level of environmental knowledge in year 2 and year 3 of undergraduate architecture students in selected private university in Klang valley.

1.1 Research in Environmental Knowledge

As mentioned in [4], demographic differences such as gender and ethnicity influenced environmental knowledge in a survey in Kentucky, United States. The study revealed that comparatively respondents from urban-metro were more knowledgeable about global issues compared to urban-non-metro and rural-non-metro respondents. The study in [5] also supported on the influence of environmental knowledge and concluded that gender and ethnicity is significance whereby blacks and women are generally less knowledgeable in terms of environmental issues compared to whites and men. As reported in [6], low level of environmental knowledge from 10th to 12th grade among high school students in national survey in United States. Environmental knowledge of children is influenced by their immediate experiences and by the content of books they use in a study of elementary school students in Greece [7].

As mentioned in [8], a significant difference in environmental knowledge levels among the three groups of science education, biology and chemistry major students in Western University, United States. A high level of environmental knowledge among their students at Michigan State University compared to the general public [9]. The research in environmental knowledge indicates significance towards

Content from this work may be used under the terms of the Creative Commons Attribution 3.0 licence. Any further distribution of this work must maintain attribution to the author(s) and the title of the work, journal citation and DOI.
Published under licence by IOP Publishing Ltd

ethnicity and gender. Urban-metro white men respondents are more knowledgeable comparatively. Environmental knowledge is also influenced by student's immediate experiences and by the content of books they use in a study of elementary school students. Environmental knowledge levels also differ among groups of science education and it is higher among educated students compared to general public.

2. Method

The 5 selected private universities with a population size of 600 architecture students from year 2 and year 3 are accredited by Lembaga Akitek Malaysia (LAM) Part 1 and Malaysia Qualification Agency [10]. Using proportion ratio method, the sample size were 234 architecture students from year 2 and year 3 from the selected 5 private universities. The pilot study with 50 questionnaires was tabulated with a Cronbach Alpha of 0.954, suggesting the items have relatively excellent reliability and internal consistency. Two PhD architecture senior lecturers validated the questionnaire before it was distributed to the study population. For the main study, descriptive analysis was used to gauge percentages mean variance and standard deviation. Since the study is parametric study, inferential statistics was used such as independence t-test.

The student survey questionnaire is divided into Section A which on demographic and background information and Section B is on level of environmental knowledge towards architectural studies. It contains 25 questions in a five-point Likert-style scale where the number "5" indicate strongly agree response and "1" indicated the strongly disagree response. The questions on Environmental Knowledge (EKn) were mapped with subjects which are related to architectural environment such as building science, green design, sustainable energy, thermal comfort, visual comfort, acoustic comfort, recycle of building materials, air pollution and water pollution. The mapping of questions is relevant to the expected knowledge in their course of study whereby students were assessed based on capability to understand and assess the impact to society on the environment. The Cronbach alpha coefficient for environmental knowledge is 0.935, suggesting that the items have relatively excellent reliability and internal consistency.

2.1 Research Design

The existing curriculum in architecture is already intensive with large credits offered in major subjects [11]. Environmental education in architecture is to increase the sustainability of the building without compromising the site and human comfort. The goal of achieving sustainability in building can be achieved with three levels of environmental education in architectural educational objectives [12]. The first level is creating environmental awareness, the second level understanding building ecosystem and the third level is the ability to design sustainable buildings as per Figure-1. The development of research questions are based on the above mentioned educational objectives which are currently pursued in architecture environmental education curriculum.

At first level, achieving sustainability in building is by instilling environmental knowledge through modules which are related to environmental science or environmental sustainable design. Students acquire knowledge to describe principles of energy conscious design, exemplify environmental strategies already adopted in the vernacular architectural language of Malaysia. Students would be able to exemplify and explain the concept of sustainability including environmental and ecological sustainability and students will be able to explain energy requirements in buildings and building materials. *Environmental Knowledge (EKn)* is information on the environment gained from a variation of experiences leading to a basic understanding of its related problems. As mentioned in [6], defined it as student's capability to understand and assess the impact of society on the ecosystem. This knowledge is established by reorganizing environmental problems as well as understanding the origins, implications and consequences of those problems. To support social groups and individuals gain a diversity of experiences in and gain a basic understanding of the environment and its associated problems according to Tbilisi Declaration [13].

Environmental Literacy	Environmental Knowledge		
Environmental Education in Architecture	capability of basic understanding	assess impact & consequences	Modules
Creating Environmental Awareness	•	•	Environmental Science
Understanding Building Ecosystems	•	•	Building Material
	•	•	Building Services
	•	•	Building Science
Ability to Design Sustainable Buildings	•	•	Environmental Design Integration

Figure 1. Environmental Knowledge Questionnaire Matrix

The second level in understanding building ecosystems is achieved by incorporating sustainability into core modules such as Building Science, Building Materials and Building Services. Thermal mechanisms such as solar radiation, heat conduction, heat resistance and heat convection are explored and how building materials respond to thermal mechanisms as well as various strategies of reducing heat gain in the interior spaces through control of building materials. Students would acquire knowledge of designing building which responds to site context and applying appropriate design strategies for climate responsive architecture. Students also would demonstrate understanding of energy conscious design by recalling and applying understanding of specific knowledge such as services systems and low embodied energy materials to be used. At this level, students also have the ability of using various laboratory equipment, tools and software to enhance their skills in applying their knowledge in designing sustainable buildings. Students would have knowledge on able to design simple acoustically-compatible spaces that enhance sound qualities in spaces while exploring noise reduction method to achieve required reverberation time and acoustic comfort.

At third level, ability to design sustainable buildings, the emphasis on design is as an integrative process. Students are required to develop environmental strategies and technological resolutions with the main intent to strengthen their design exploration/ideas, and complement their design through a comprehensive sustainable environmental design. There are 6 accepted principles that constitutes sustainable which are: i) optimize site potential, ii) optimize energy use, iii) protect and conserve water, iv) use environmentally sustainable products, v) enhance indoor environmental quality and vi) optimize operational and maintenance practices [14]. In line with this, Malaysian Green Building Index which conforms to MS1525:2007 also emphasizes the above principles.

It is imperative for student's design to be conforming to building performance for instance thermal comfort, lighting, acoustic, conservation of water, appropriate building services systems, building material such as green materials, sustainable technology and construction methods while adhering in principle to regulatory uniform building by-laws (UBBL) and BOMBA fire regulations in order to achieve LAM recognition of Part 1. To prevent these shortcomings in Malaysia, it is critical for the studio work to put emphasis on integration of sustainable objectives together with other learning outcomes.

3. Results

3.1 Descriptive Information of the Population Size, Sample Size, Year of Study, Gender and Age

The population size (N) of year 2 and year 3 undergraduate architecture students in Institution 1 is 42 students. The population size of year 2 and year 3 students in Institution 2 is 50 students. The population size of year 2 and year 3 undergraduate architecture students in Institution 3 is 48 students. The population size of year 2 and year 3 students in Institution 4 is 49 students. The population size of year 2 and year 3 undergraduate architecture students in Institution 5 is 413 students. The total undergraduate year 2 and year 3 architecture students in all the selected universities are 602. For each selected university, the sample size (S) is extracted based on proportion ratio method using the formula below.

$$\frac{\text{Total Architecture students in one University}}{\text{Total Population Size of All Universities}} \times \text{Total Sample Size}$$

The number of sample size is based on universities and average mean age is well distributed among the universities. The total sample size for year 2 is 114 (48.72%) and total sample size of year 3 is 120 (51.28%). Year 2 and year 3 undergraduate architecture students age ranges from 20 to 23 years old. The mean age of student's sample size is 21.12. The total male sample size is 124 (53.00%) and the female sample size is 110 (47.00%). The selected universities have equal sample size number of year 2 and year 3 undergraduate architecture students and equal well distributed male and female undergraduate architecture students.

3.2 Descriptive Analysis using Mean Score for Environmental Knowledge Variables in Year 2, Year 3 and Combine Year 2 & Year3

To analyse the level of environmental knowledge in year 2 and year 3 of undergraduate architecture students in private universities, the mean score and standard deviations for year 2 and year 3 are measured separately. The level can be determined by using mean score and Table 1 discusses the interpretations of mean scores to determine the environmental knowledge variables level for year 2, year 3 and combine year 2 & year 3 in selected private university.

Table-1. Mean Score Interpretation [15]

Mean Score	Level
1.00 to 2.33	Low
2.34 to 3.66	Medium
3.67 to 5.00	High

The results show that the environmental knowledge level is high with mean score (*M*=4.19, *SD*=.428) for year 2 architecture students in selected private universities in Klang valley. For year 3 students, environmental knowledge level is also high with mean score (*M*=4.09, *SD*=.569). For combine year 2 and year 3, environmental knowledge level is also high with mean score (*M*=4.14, *SD*=.506). Based on independent sample t-test, there are no significant mean differences for variables of environmental knowledge between year 2 and year 3 from selected private universities Environmental Knowledge (EKn), t (220) =1.353, p=0.177. These results suggest that environmental knowledge variables do not have influencing factor in years of study namely for Year 2 and Year 3 from selected private universities.

4. Discussion

The study reveals a high level of literacy in environmental knowledge in year 2, year 3 and combination of year 2 and 3 students. The results indicate that the student's information on the environment gained from a variety of involvements to a basic understanding of its related problems is also at high level. Student's environmental knowledge is demonstrated by reorganizing environmental issues as well as grasping the roots, implications and consequences of those problems. Student's learning outcome in the environmental subjects which are designed in the current context for undergraduate architecture students are more relevance towards environmental knowledge which is in line with Blooms Taxonomy educational objectives. This study is in congruent with [7], observed that student's environmental knowledge is influenced by their immediate involvements and by the content of books they use in Greece. Surprisingly, earlier findings [6] reported low level of environmental knowledge among high school students in national survey in United States.

5. Conclusions

The results suggest that high level of environmental knowledge in year 2, year 3 and combination of year 2 and year 3 students. This supports the Malaysian architectural education accreditation body which sets "a satisfactory balance between theory and practice" as its first qualitative standards, in terms of the anticipated scope of competencies that an architecture student needs to attain and identifies design capabilities, knowledge, and skills to accomplish an architect's role as generalists who can co-ordinate interdisciplinary objectives. The council also endorses teaching based on project realization as the principle teaching method. This approach places students under the direct and personal guidance of lecturers. This method provides "a synthesis of knowledge, aptitudes, and attitudes" [3]

Acknowledgement

The author would like to acknowledge everyone from every universities especially from the architecture department who assist and gave precious inspiration to the study.

References

[1] Orr D W (2005) Ecological literacy: Educating our children for a sustainable world Sierra Club Books

[2] Zalina Shari and Mohamad Fakri Zaky Jaafar 2006 Towards a holistic sustainable architectural education in Malaysia *International Journal on Sustainable Tropical Design Research & Practice* 1(1) 57–65

[3] Lembaga Arkitek Malayisa (LAM) 2015a Policy and procedure for accreditation of architectural programmes. Retrieved September 29 2015 from http://www.lam.gov.my/index.php/accreditation/interior-design-policy-procedure.html

[4] Arcury T A and Christianson E H 1993 Rural-urban differences in environmental knowledge and actions *The Journal of Environmental Education* 25(1) 19–25

[5] Zimmerman L K 1996 Knowledge affect and the environment 15 years of research *The Journal of Environmental Education* 27(3) 41–44

[6] Gambro J S and Switzky H M 1996 A national survey of high school student's knowledge *The Journal of Environmental Education* 27(3)

[7] Paraskevopoulos S, Padeliadu S and Zafiropoulos K 1998 Environmental knowledge of elementary school students in Greece *The Journal of Environmental Education* 29(3) 55–60

[8] Robinson M and Crowther D 2001 Environmental science literacy in science education biology and chemistry majors *The American Biology Teacher* 63(1) 9–14

[9] Kaplowitz M D and Levine R 2005 How environmental knowledge measures up at a big ten university *Environmental Education Research* 11(2) 143–160

[10] Malaysian Qualifications Agency (MQA) 2015 Malaysian qualifications register Retrieved April 1 2015 from http://www.mqa.gov.my/mqr/

[11] Committee on the Environment (COTE) 2007 Ecological literacy in architecture education Retrieved from http://www.aia.org/practicing/groups/kc/AIAS074665

[12] Kim J J, Rigdon B and Graves J 1998 Introductory module pollution prevention in architecture national pollution prevention centre for higher education Michigan University of Michigan

IOP Conf. Series: Materials Science and Engineering **291** (2017) 012010 doi:10.1088/1757-899X/291/1/012010

[13] United Nations Educational Scientific and Cultural Organization and United Nations Environment Programme (UNEP) 1977 Tiblisi Conference Declaration United Nations Education Scientific and Cultural Organization Retrieved from http://www.gdrc.org/uem/ee/tbilisi.html
[14] Whole Building Design Guide (WBDG) 2007 Whole building design guide sustainable committee Retrieved from http://www.wbdg.org/design/env_preferable_products.php
[15] Polit D F and Hungler B P 1999 Nursing Research Principles and Methods sixth ed J B Lippincott Company Philadelphia New York Baltimore

Client-oriented Building Mass Customization (CoBMC)

Xia Sheng Lee, Mohd Faris Khamidi, Sivaraman Kuppusamy and Chin Tuck Heng

School of the Built Environment, University of Reading Malaysia, Persiaran Graduan Kota Ilmu, Educity 79200 Iskandar Puteri Johor, Malaysia

x.s.lee@reading.edu.my, leexiasheng@gmail.com

Abstract. Although much later compared to other industries including aerospace, automobile, oil and gas etc., digital technology development has been cresting towards an exponential curve in the construction industry. Technological diversity and abundance change the game from "what you can" to "what you want". Society is changing at an unprecedented rate. Consequently adaptability will be crucial. This research paper explores the integration of digital adaptive technologies that transform the construction industry from the mass production to that of a possible client-oriented mass customization. The focus on the design, construction and performance stages of a building project, currently undergoing major overhaul faces a paradigm shift globally that will impact and compel attention for the next three decades with viable solutions such as Building Information Modelling (BIM) to manage massive data cum information. Customization maximizes clients' participation during the design process thereby achieving greater effective value and higher satisfaction. A study between customized and standardized examples will investigate how adaptive customization will shift the design paradigm from cost to value centric. This action research will explore different aspects of emerging innovative systems already in place pushing the edge of frontiers, and transforming the building industry landscape whether micro or giga, to compliment new technologies to create an unprecedented exhilaration of freshness over the mundane, routine and mediocrity. Three identified fundamental aspects that are instrumental to Client-oriented Building Mass Customization (CoBMC) are design option visualization, parametric product information and n-dimensional modelling. The study concluded that a paradigm shift is therefore inevitable for every stakeholder including clients who will need to re-examine their roles, capabilities, and competencies in preparation towards challenging future.

1. Introduction

Client or customer satisfaction is always the ultimate factor for business success. This is no exception for the construction industry as purported by The Latham Report of 1994, which specifically recognized clients as the driving force: "Implementation begins with clients. Clients are at the core of the process and their needs must be met by the industry. But clients are dispersed and vary greatly" [1]. Every client certainly will have his or her own distinct requirements, specific concerns and unique preferences. Is it possible to create a typical planning, design and construction that fits all, while achieving excellent client-satisfaction [2] [3] at the same time? Customization is a useful tool to satisfy client requirements that is unique in the construction industry. Whilst customization empowers clients to make micro decisions that will trim the gap between their expectation and the final construction outcome, this will

Content from this work may be used under the terms of the Creative Commons Attribution 3.0 licence. Any further distribution of this work must maintain attribution to the author(s) and the title of the work, journal citation and DOI.
Published under licence by IOP Publishing Ltd

IOP Conf. Series: Materials Science and Engineering **291** (2017) 012011 doi:10.1088/1757-899X/291/1/012011

lead to the "I designed it myself" effect [4] and directly increase client satisfaction. The idea of custom-made buildings is not new.

It is of intrigue to look at speculations in relation to how the construction and housing industry evolved since the year 2000:

"By 2015, savvy well-capitalized companies from outside of the housing industry have taken over the market, and speculative developers have all but vanished. A young couple looking to build a new home begins the process at one of a number of internet home sites, where they play design games and select from options presented to them. As the system begins to understand their needs and values, proposals are produced for their evaluation - which then, in turn, leads to additional designs. In the process, they find that there has been an explosion of creative activity as dozens of young architects and industrial designers develop and license systems to integrators - based on new industry-wide standards for both physical building components and digital technologies."
- Larson, K. [5]

Obviously, such speculations did not materialize, at least not on a massive scale. This is possibly due to factors such as evolving technology, statutory regulations, lack of skilled human resources, escalating cost and stakeholders' perception, etc. However, continuous research towards custom-made buildings has not stopped by exploring through the means of integrated system [6] and information technology [7]. As Building Information Modelling (BIM) [8] implementation matures, there is great potential in realizing mass customization of the built environment in an unprecedented scale. This research is a conceptual study about the possible Client-oriented Building Mass Customization (CoBMC) via BIM or Virtual Design and Construction (VDC).

1.1. Client-Oriented Process
There are many labels such as "client", "customer", "service user" or "consumer" depending on business types and disciplines, which describe not of a person, but of inter-relationships [9]. It is a widely accepted fact that a client-oriented approach through collaboration [3] and integration [6][10] that will strengthen the client relationship. Non-construction products have ventured into "design by customer" interaction [11] achieving the next level of client-oriented processes. Customization allows personalization and innovation which are not achievable with mass production. As client requirements and construction industry expectations broaden and deepen, creative approaches will inevitably and earnestly be adopted to enhance client-oriented processes. Buildings are one of the most expensive products one would purchase and own for a relatively long period of time. It is only logical that clients of the construction industry have access to a mass customization process to meet their own needs.

1.2. Building Mass Customization with Building Information Modelling
There are many labels such as "client", "customer", "service user" or "consumer" depending on business types and disciplines, which describe not of a person, but of inter-relationships [9]. It is a widely accepted fact that a client-oriented approach through collaboration [3] and integration [6][10] that will strengthen the client relationship. Non-construction products have ventured into "design by customer" interaction [11] achieving the next level of client-oriented processes. Customization allows personalization and innovation which are not achievable with mass production. As client-requirements and construction industry expectations broaden and deepen, creative approaches will inevitably and earnestly be adopted to enhance client-oriented processes. Buildings are one of the most expensive products one would purchase and own for a relatively long period of time. It is only logical that clients of the construction industry have access to a mass customization process to meet their own needs.

Mass customization for building refers to both the number of buildings being customized and number of customizations being made in a single building. Earlier study on parametric space configuration, for example, the Palladian grammar [12] is significant to later building mass customization research like Siza's houses at Malagueira[13]. After that, computer systems were developed integrating 3-dimensional computer aided design (CAD) and spreadsheets to facilitate [6] building design. Such computerized framework required the designer to create a system, rather than a specific design. However, this is time consuming and costly when compared to traditional approaches. The balance between the customizable flexibility level and customization complexity will be a determinant to achieving successful building mass customization. As technology advances, the obstacles towards a practical building mass customization solution will be collaterally a greatly transformed entity.

Building Information Modeling (BIM) is about an integrated approach to design, construction, and facility management [14]. Such an approach is considered "new" when compared to the traditionally perceived 2-dimensional Computer Aided Design (CAD). BIM framework is introduced later to better understand and integrate the relevant fields such technology, process and policy [15]. It also spells out "what is" and "what is not" BIM by clearly defining the stages namely, object-based modelling, model-based collaboration, and network-based integration. BIM adoption and implementation will be easier by deliberately developing all the matrix components with a holistic approach [16]. Research also shows that majority of construction industry stakeholders agree that greater use of BIM will improve construction best practices [17]. Powerful information systems [18] will manage the increase of information [19] that supports the growth of customization heterogeneity. BIM offers an efficient way to integrate the massive information [20] generated for optimization [21] during the mass customization process.

2. Methodology

This is a conceptual action research [22] type study with no collection of hard data, but BIM model outputs for examination and exploration. Action research is known for uncertainty and complexity, which is suitable to explore evolving research directions. The study collaborates with highly experienced practitioners [23] in the building industry including architects, engineers and quantity surveyors. The study explores and predicts how the building industry will transform from industry-oriented mass production to client-oriented mass customization from various significant aspects.

3. Significant Aspects Of Client-Oriented Building Mass Customization (CoBMC)

There are a few aspects that will make mass customization more feasible than before by using BIM, namely design option visualization, digital product information and n-Dimensional modelling. Some may argue that such aspects are available even without BIM. However, the advancement in computer hardware and software capabilities has critically enhanced the quality and quantity of these aspects. The first aspect is design option visualization. Without BIM, the number of design alternatives is limited, however, it is now feasible to have hundreds of design alternatives explored and visualized once it has been set up. The second aspect is BIM provides parametric product information for almost every types of building components thereby making actual construction true to its intentions. The third aspect is n-dimensional modelling that integrates the information of time, cost and quality for efficient decision-making.

3.1. Client-Oriented Building Design Process

Written project brief and 2D layouts certainly are essential to communicate with clients. However, 3D visualization is the key for mass customization as it enables clients to see promptly instead of imagining the construction outcome. Design option visualization is enhanced by the 3D modelling efficiency which reduced the gap between imagination and visualization as shown in Figure 1 and Figure 2 This action research learned that even on the same structural system, design option visualization makes a lot of difference in communication with the client.

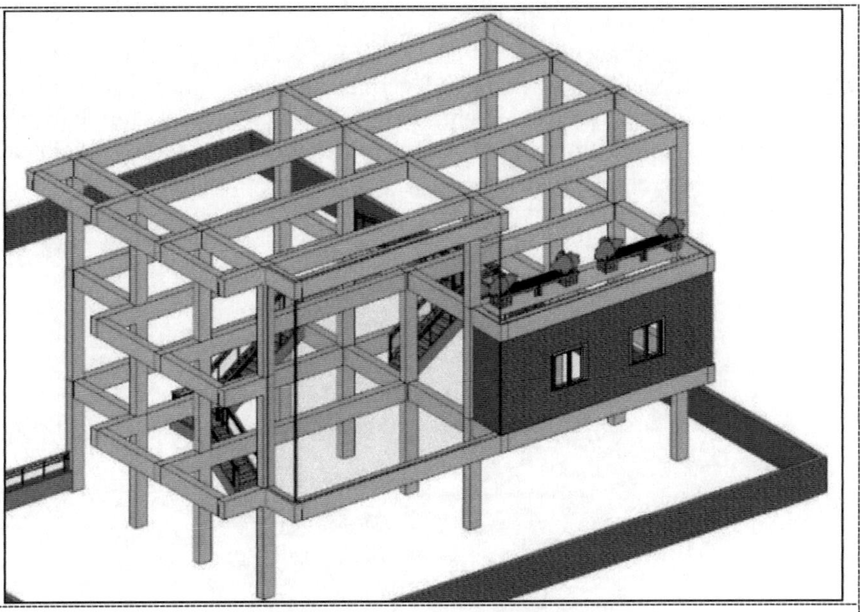

Figure 1. Design option A

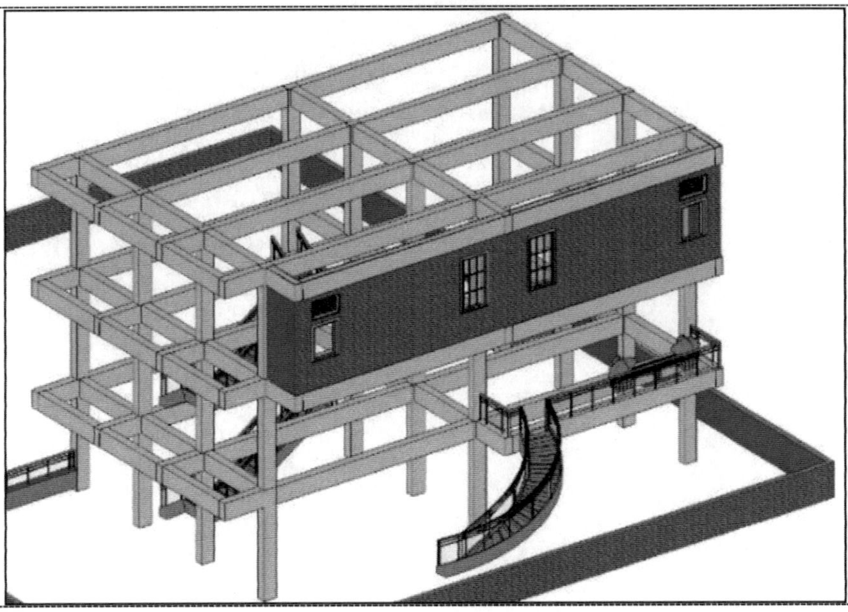

Figure 2. Design option B

Design option A as shown in Figure 1 is designed based on the client's requests for a transparent double volume living room, playroom with good sound insulation at level 1, and some green plants at level 2. Design option B as shown in Figure 2 is a variation when the client requested for a sheltered balcony, an external curved staircase going up to level 1, and an enlarged master bedroom at level 2. Client's requirements are being scrutinized at the design stage through various high level considerations. Mass customization with design option visualization will increase client's satisfaction, thereby encouraging green construction by reducing waste, rework and renovation after initial construction.

3.2. Parametric Product Information

Parametric product information provided by manufacturers is a significant enabler for mass-customization, as now designers can "plug and play" with existing building products including finishes and fittings such as, facade panel, furniture, light fittings, sanitary fittings, etc. This action research exercise found that mass-customization can be done by exploring numerous combinations of two different parametric products as shown in Figure 3.

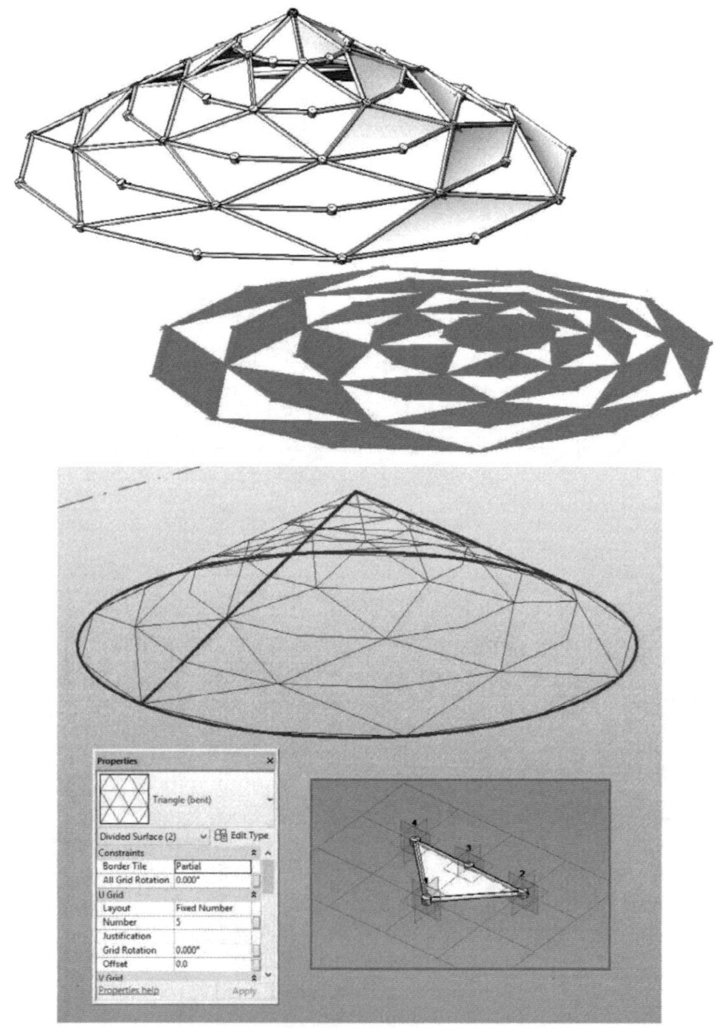

Figure 3. Parametric product information

Figure 3 shows that the two parametric products namely (1) umbrella shaped steel structure, and (2) curtain panel, under mass customization with the combination of both products, the variance in height, width, thickness, material, and colour etc., clients can customize the final outcome to their unique preference instead of the compulsion to select mass-produced elements. The example demonstrates the "I designed it myself" effect that consequentially enhanced client's satisfaction.

3.3. N-Dimensional Modelling

Management tools such as the Gantt chart and the S Curve chart are useful to visualize and correlate different information such as cost and time. BIM's n-dimensional modelling drastically enriched such tools to highlight the impact of various customized designs in real-time to the clients for pertinent decision-making as shown in Figure 4, which able to identify the detail information of every single curtain panel in the building.

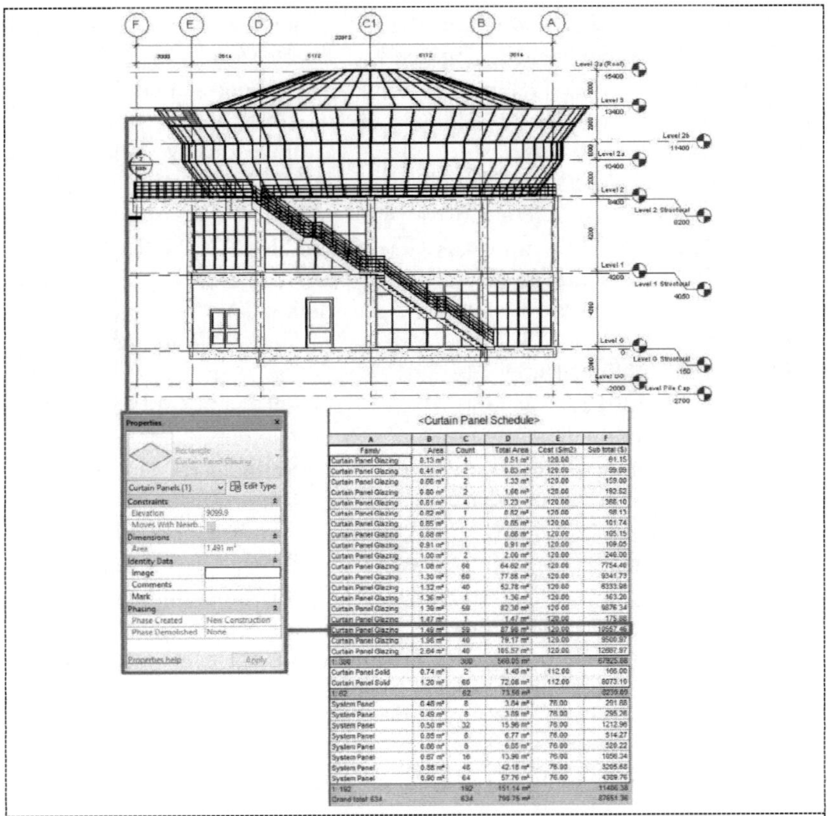

Figure 4. Information of different dimensions in an integrated model

This action research found that by utilizing the function of automated quantity extraction from n-dimensional model shown in Figure 4, clients are able to participate proactively even in the most detail stage of the design. Any impact such as surge in cost or time delay communicated graphically to the client to suit personal or organizational considerations is interactive. On-going changes are adeptly automated to cater for prompt updates.

4. Conclusion

Clients' requirement and expectation are becoming higher over the years, and this is an irreversible trend. A paradigm shift is therefore inevitable for every stakeholder including clients who will need to re-examine their roles, capabilities, and competencies in preparation towards challenging times. This action research presents a fresh approach for building mass customization by featuring the significant aspects including design option visualization, digital product information and n-dimensional modelling to spearhead a much-needed change. These aspects are instrumental for any building research, innovation and development from not just the technological perspective, but also address the client's behavior. The research is timely to explore the transformation for building development from industry-oriented mass production to client-oriented building mass customization to infuse personality and preferences in an ever-depersonalized world. Further research shall investigate the compounding effects of other technological advancements within the Client-oriented Building Mass Customization such as 3D printing, computer numerical control (CNC), reverse engineering, cloud computing etc.

References

[1] Latham, M. Sir. (1994). The Latham Report: Constructing the Team. Retrieved October 2015, from http://www.cewales.org.uk/cew/wp-content/uploads/Constructing-the-team-The-Latham-Report.pdf

[2] Ahmed, S. M., & Kangari, R. (1995). Analysis of client-satisfaction factors in construction industry. Journal of Management in Engineering, 11(2), 36-44.

[3] Bresnen, M., & Marshall, N. (1999). Achieving customer satisfaction? Client-contractor collaboration in the UK construction industry. In Customer Satisfaction: A Focus on Research and Practice, Joint Triennial Symposium of CIB Commissions W (Vol. 55, p. W65).

[4] Franke, N., Schreier, M., & Kaiser, U. (2010). The "I designed it myself" effect in mass customization. Management Science, 56(1), 125-140.

[5] Larson, K. (2000). The home of the future. ARCHITECTURE AND URBANISM-TOKYO-, (10; ISSU 361), 60-67. Retrieved October 2015, from http://eecs.wsu.edu/~holder/courses/cse6362/pubs/larson00.pdf

[6] Benros, D., & Duarte, J. P. (2009). An integrated system for providing mass customized housing. Automation in Construction, 18(3), 310-320.

[7] Lee, S. K., & Yu, J. H. (2012). Success model of project management information system in construction. Automation in construction, 25, 82-93.

[8] Jung, Y., & Joo, M. (2011). Building information modelling (BIM) framework for practical implementation. Automation in Construction, 20(2), 126-133.

[9] Delgado, F., Martínez, R., Puche, J., & Finat, J. (2015). Towards a client-oriented integration of construction processes and building GIS systems. Computers in Industry, 73, 51-68.

[10] Delgado, F., Martínez, R., Puche, J., & Finat, J. (2015). Towards a client-oriented integration of construction processes and building GIS systems. Computers in Industry, 73, 51-68.

[11] Tseng, M. M., & Du, X. (1998). Design by customers for mass customization products. CIRP Annals-Manufacturing Technology, 47(1), 103-106.

[12] Stiny, G., & Mitchell, W. J. (1978). The palladian grammar. Environment and planning B, 5(1), 5-18.

[13] Duarte, J. P. (2005). Towards the mass customization of housing: the grammar of Siza's houses at Malagueira. Environment and planning B: Planning and Design, 32(3), 347-380.

[14] Eastman, C., Eastman, C. M., Teicholz, P., & Sacks, R. (2011). BIM handbook: A guide to building information modeling for owners, managers, designers, engineers and contractors. John Wiley & Sons.

[15] Succar, B. (2009). Building information modelling framework: A research and delivery foundation for industry stakeholders. Automation in construction, 18(3), 357-375.

[16] Succar, B., & Kassem, M. (2015). Macro-BIM adoption: Conceptual structures. Automation in Construction, 57, 64-79.

[17] Elmualim, A., & Gilder, J. (2014). BIM: innovation in design management, influence and challenges of implementation. Architectural Engineering and design management, 10(3-4), 183-199.

[18] Blecker, T. (2005). Information and management systems for product customization (Vol. 7). Springer Science & Business Media.

[19] Jarvenpaa, S. L., & Ives, B. (1994). The global network organization of the future: Information management opportunities and challenges. Journal of management information systems, 10(4), 25-57.

[20] Lee, X. S., Tsong, C. W., & Khamidi, M. F. (2016). 5D Building Information Modelling–A Practicability Review. In MATEC Web of Conferences (Vol. 66, p. 00026). EDP Sciences.

[21] Lee, X. S., Yan, C. P., & See, Z. S. (2016). Irregular shaped building design optimization with building information modelling. In MATEC Web of Conferences (Vol. 66, p. 00027). EDP Sciences.

IOP Conf. Series: Materials Science and Engineering **291** (2017) 012011 doi:10.1088/1757-899X/291/1/012011

[22] Brydon-Miller, M., Greenwood, D., & Maguire, P. (2003). Why action research? Action research, 1(1), 9-28.
[23] Huang, H. B. (2010). What is good action research. Action Research, 8(1), 93-109.

A structural regression model for relationship between indoor air quality with dissatisfaction of occupants in education environment

Hamid Reza Hosseini[1], Mohd Yazid Mohd Yunos[1], Sumarni Ismail[1], Maheran Yaman[2]

[1]Department of Landscape Architecture, Faculty of Design and Architecture, University Putra Malaysia, 43400 UPM Serdang, Selangor, Malaysia

[2]Department of Landscape Architecture, International Islamic University.

Hamidrezahosseini2013@gmail.com

Abstract. This paper analysis the effects of indoor air elements on the dissatisfaction of occupants in education of environments. Tries to find the equation model for increasing the comprehension about these affects and optimizes satisfaction of occupants about indoor environment. Subsequently, increase performance of students, lecturers and staffs. As the method, a satisfaction questionnaire (SQ) and measuring environment elements (MEE) was conducted, 143 respondents at five classrooms, four staff rooms and five lectures rooms were considered. Temperature, air velocity and humidity (TVH) were used as independent variables and dissatisfaction as dependent variable. The hypothesis was tested for significant relationship between variables, and analysis was applied. Results found that indoor air quality presents direct effects on dissatisfaction of occupants and indirect effects on performance and the highest effects fallowed by temperature. These results may help to optimize the quality of efficiency and effectiveness in education environments.

Keywords: dissatisfaction, indoor air quality, temperature, humidity, air velocity, education environment.

1.Introduction

Long empirical studies have considered the impact of components of indoor environment on residential. But a few of them have investigated the impact of individual indoor air quality on occupants' perception, satisfaction and performance in education environments. Classroom is where students spend a lot of time of day in there with high density of pupils which causes to aggravate poorer indoor air quality. Furthermore, education environment can be a place where are required multiply function or duties. These duties were categorized in two main parts practically and mentally [1]. Analysis of indoor air quality factors, such as temperature, humidity and air velocity in education environment can help to improve learning and teaching [2]. Base on literature review many factors interacted in dissatisfaction of occupants like lighting, acoustics, furnishing, technology, indoor air quality, thermal comfort and so on. But, this paper focuses on temperature, humidity, air velocity. Because many researchers stated (TVH) are more important than other elements [3]. The purpose of this study is designing the equation model with statistically validated that regard the relationship between (TVH) and dissatisfaction. This paper won't explore the effects of other environment elements. It is widely accepted that air conditioning is associated

Content from this work may be used under the terms of the Creative Commons Attribution 3.0 licence. Any further distribution of this work must maintain attribution to the author(s) and the title of the work, journal citation and DOI.
Published under licence by IOP Publishing Ltd

with absentee rates or academic performance and indoor air quality has significant benefits to health and performance. In additional, the human right emphasis on healthy indoor environment which includes the right to breathe clean air, the right of visual health, comfort and the right of the thermal comfort[4].

2. Literature review

By using keywords, such as temperature, air velocity, humidity, satisfaction of occupants and performance, the literature was reviewed in several databases (Google scholar, Elsevier, Science direct) to explore the presence of the indoor air factors in predicted dissatisfaction of occupants in education environments. Form 1960's till now a large quality researches have been done in this field. Maybe, one of the first one was McNall & Nevins in 1967 that by thermal environment control in or near the comfort zone indicated a strong non- statistically significant tendency is between academic performances with the air- conditioned in junior school student in1959[5]. However, this hypothesis couldn't determine whether these changes exactly related to air- condition or other variables like student motivation, preparation of student, back ground or teachers. Schoer and Shaffran (1973) conducted an experiment in two classrooms which temperature and relative humidity could be controlled but in one air conditioned and just ventilation used. The result of the coursework and tests were identified complex tasks were affected more than simple learning [6]. One of the most important factors in evaluation poor indoor air quality is absenteeism of worker and student. Green (1974) evaluated the absenteeism as function of the humidity. The study has been found that increasing relative humidity causes to reduce illnesses and respectively absenteeism will reduce [7].Mclntyre(1980) has studied about the reaction of people to air velocity and temperature. In this study found that the power of cooling air is more effective than the speed of air and mention to the standards which the German express (air velocity 0.18 m/s with 22^0c) and (0.30m/s with 27^0c). And finally, Mclntyre refers to the mean of 0.3m/s for air velocity with ±0.05 that cause maximum of discomfort for highly occupied spaces [8].Fanger (1988) describes two units in terms of evaluation of air quality which are the Olf and the Decipol [9]. Myhrvold (1996) during three-years in Norwegian school carried out to estimation of the relation between concentrations of carbon dioxide with student performance. The results shown performance were reduced by carbon dioxide levels because of poorer indoor air quality [9].Clements-Croome & Robert (2008) explained how the air passes via nose to olfactory systems and consequence to the respiratory system where smoke, dust, radon and formaldehyde besides nitrogen and oxygen are transferred through the lung to the blood and hence is circulated around body. In this study has presented a new definition of the word "fresh" that can be associated with cold, cleanliness, new, preserved, salubrious and ventilated. Croome & Robert and Purecll & thorne in 1987 in another papers determined how freshness are evaluated by parameters like temperature, air velocity, relative humidity and they built the theory which namely arousal level theory. This theory base on people responded about freshness and founded the significant relationship between temperature fresh airs supplies with freshness [10]. Fisk and Rosenfeld (1998) have stated that due to reduction of respiratory infection productivity increase [11]. Seppanen et al (2006) has estimated a quantitative relationship between productivity and performance with ventilation for working in office[12]. Wood and Skaret (1999) has identified to improve indoor air quality because increased productivity that their profit is 10-100 times greater than the cast of maintenance, annualized construction and building energy. In 1990, M. Amerigo and J.Aragones, presented a conceptual framework for residential satisfaction. This model implies a dynamic interaction between occupants' satisfaction and their environment. In this systematic model are replaced cognitive and behavioral affective as factors which affected direct or indirect on satisfaction of occupants [13]. Humphreys MA (2005) stats that human health, productivity, stresses and wellbeing always influenced by the quality of the indoor environment. This paper is noted to the hypothesis that overall satisfaction of occupants related to way in which they perceived indoor environment. Many researchers are demonstrated that this hypothesis is correct [14]. Frontczak and Wargocki (2011) by analyzing surveys which sent to office building found that the main indoor environment parameters that impact on office occupants are temperature, air quality, light, noise level and sound privacy [15]. In 2009, lai JHK et al. demonstrated that thermal comfort has the highest impact on overall satisfaction in commercial buildings. And, this was followed by noise and air quality [16]. Another investigation in office building showed that ventilation could also have great impact on overall satisfaction of occupants (Steemers K, Manchand S 2010) [17]. The relationship between thermal comfort, air quality, sound quality and overall satisfaction has been widely investigated by A.Zalejska-Jonson, M. Wilhelmosson(2013)[18]. Base on the literature review of this past research, the results obtained to show that indoor air quality has adverse health effects and losing performance. In this field many researches have been done about the impacts of indoor environment quality with productivity and performance in office buildings but less studies considering to indoor air quality in academic environments. However, environments conditions in schools and universities much worse than office buildings. In addition, students have to do mentally and perform work that is not optional and is almost new to them also they cannot change their job to the healthy worker.

3. Methodology

air conditions and windows. The measurements were done by HD32.3- Delta OHM. Each reading data took average within fifteen minutes. This device has three sensors, namely air temperature The methodology applied in this paper comprises the following stapes: 1- objective measurement (temperature, air velocity and humidity) by device. 2- Subjective measurement (Conducting and application of the questionnaire with themes of satisfaction and perceived). 3- Statistical analysis. 4- Estimate the regression equation model and test it. Objective measurements: The indoor thermal investigation consisted of air temperature that included three items 1-Globe thermometer temperature 2- Ambient temperature 3- Radiant medium temperature, relative humidity, air velocity and Predicted Percentage of Dissatisfied. Air conditions were switched on during measurement. Windows were closed throughout the measurement period. In each room or classroom, the measurements had done three times in different places of the room and different distances from, relative humidity and air velocity. Subjective measurements: The subjective measurement was conducted using questionnaire survey. To aid subjects, the questionnaire was prepared with Bahasa Malaysia translation and glossary of sick building syndrome, satisfaction, efficiency. Students, staff, and lecturers were approached. The restriction in the sample collection was due to students that spend a whole day in the class like the students in the studio of design. They were asked for their gender, age, and type of relationship with this faculty. This questionnaire survey conducted in four sections. Section A is about personal. Section B sick building syndrome with Section C, satisfaction and perceived section D efficiency and effectiveness. For all parts used a liker scale of seven answers that were written a pattern of each question as a guide. The sample size is included 143 participations. The sample size was $n>30$ according to the Central Limit Theorem it has to show the distribution of the people and chosen randomly .Statistical analysis: In this section were used the statistical methods and graphs for comparing the data. The mean, sample standard deviation, the standard error for a sample mean and linear Regression explained the data obtain from device and questionnaire.

4. Data and analysis

As mention before, we have three main Independents variables which are temperature, related humidity and air velocity and one dependent variable predicted percent dissatisfaction. First of all, correlation will be estimated for each of them with satisfaction. Then, regression equation will be estimated at multivariate. Finally, it will test the significance of a regression equation to determine whether the sample correlation represents areal relationship or is simply the result of sampling error. For the first step Temperature, humidity, air velocity and dissatisfaction have been ranked and then computed the correlation between two variables. The analysis examined the relationships among (TVH) and satisfaction from indoor air quality for n=143 participants. Comparing data showed a consistently one –tail directional relationship but not necessarily a linear relationship. For this reason the (rank-order) spearman correlation used for this experiment and evaluated by hypothesis.

H_0: $\rho=0$ the null hypothesis states that there is "NO" correlation between these variables or the relationship between temperature/ humidity/air velocity and dissatisfaction is equal zero.
H_1: $\rho\neq0$ the alternative hypothesis states that there is a real relationship between temperature/humidity/air velocity and satisfaction.

data	Sub-objective	correlation	n	α	df	type	table	Sig/non Sig	H_0/H_1	r^2
1-Survey	temperature AM-DISSAT	0.192952001	143	$\rho<0.05$	141	One-tail	0.164	Significant*	Reject H_0	0.037230475
2-Survey	Temperature PM-DISSAT	0.620740175	143	$\rho<0.05$	141	One-tail	0.164	Significant**	Reject H_0	0.385318365
3-Survey	Humidity-DISSAT	0.167737779	143	$\rho<0.05$	141	One-tail	0.164	Significant*	Reject H_0	0.028135962
4-Survey	Air velocity-DISSAT	-0.09524278	143	$\rho>0.05$	141	Two-tail	±0.194	Non-significant	fail reject H_0	0.009071188
5-Device	Temperature-DISSAT	0.810305356	108	$\rho<0.05$	106	Two-tail	±0.194	Significant***	Reject H_0	0.65659477
6-Device	Humidity-DISSAT	0.105247179	108	$\rho>0.05$	106	Two-tail	±0.194	Non-significant	Fail reject H_0	0.011076969
7-Device	Air velocity-DISSAT	0.139518857	108	$\rho>0.05$	106	Two-tail	±0.194	Non-significant	Fail reject H_0	0.019465512

NO	Structural equation model	$Y=SAT, X_1=T(AM), X_2=T(PM), X_3=HU, X_4=AV$	F-ratio
8-	Temperature(AM) & dissatisfaction	$Y= 0.192952001\ X_1+ 58.10745592$	5.452495
9-	Temperature(PM) & dissatisfaction	$Y= 0.620740175\ X_2+ 27.30670738$	88.38703
10-	Humidity & dissatisfaction	$Y= 0.167737779\ X_3+ 59.92287994$	4.082022
11	Air velocity & dissatisfaction	$Y= -0.095242785\ X_4+ 78.85748055$	1.290746
12	T(PM),HU,AV& satisfaction	$Y=0.646950317X_2-0.045286726X_3-0.118574031\ X_4+ 37.21755175$	31.57514

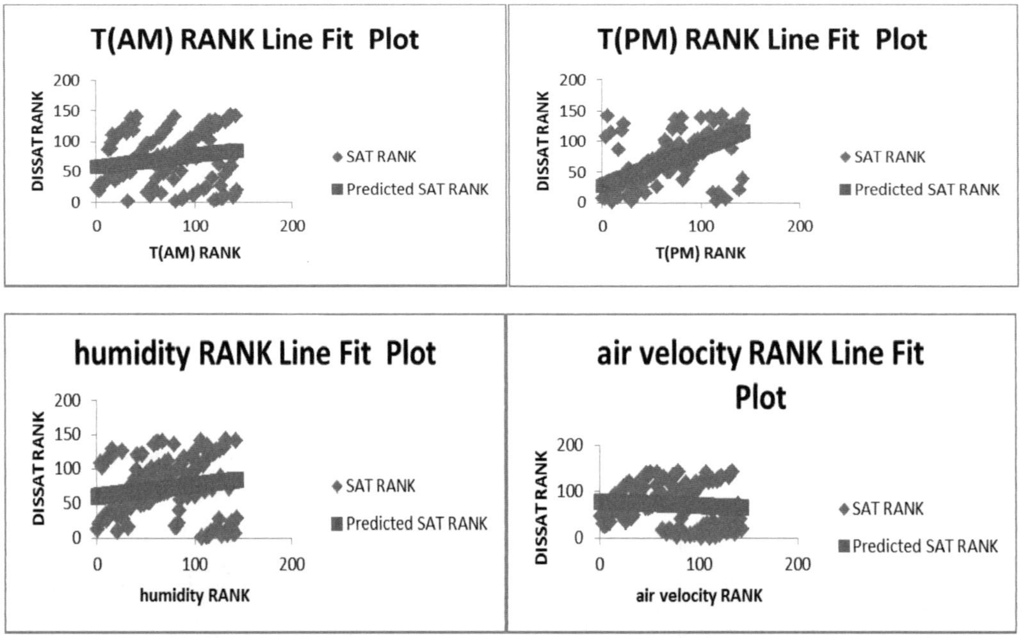

NO	Structural equation model	Y=dissatisfaction, X_1=T, X_2=HU, X_3=AV	F-ratio
13	T, HU, AV & dissatisfaction	$Y= 8.874517402\ X_1 + 0.009356889\ X_2 - 2.018576993\ X_3 - 191.6488083$	66.76176716

5. Results

By analyzing surveys and device data found that the main indoor air parameters that impact on dissatisfaction of occupants in education environments is temperature. The vote of participants and device shown that between 38 till 65 percentage of dissatisfaction of occupants predicted by temperature. But, less than 2 percentage of dissatisfaction predicted by humidity and air velocity. In the other words, temperature interference is more than other aspects in dissatisfaction. In number 12 and 13 in table above we have two regression equation models by survey and device, they predicted dissatisfaction of occupants by temperature, humidity and air velocity. The R squared is 0.405285079, 0.658215499. Therefore, dissatisfaction will be predicted between 40 to 65 percentages by these variables or these regression models. The variance predicted by regression equation is significantly greater than would be expected if there is no relationship between variables. In the other words, the null hypothesis will reject and the regression equation predict significant portion of the variance of the dissatisfaction. As mention above, one of the objectives of this paper is increase performance in education environments, so, the model from device is better predictor than model (12) for dissatisfaction. In the other words, the equation model number (13) can predict the level of dissatisfaction of occupants. And, this level has indirectly relationship with performance. It means by increasing dissatisfaction performance of residential in education environment will decrease. Base on the literature review, there is the same relationship between dissatisfaction and ventilation rate [19]. Consequence, if dissatisfaction is replaced in equation between predict dissatisfaction and ventilation rate; therefore, the ventilation rate that necessary for improve indoor air quality will be estimated. Or the ventilation rate for decreasing dissatisfaction of residential will be estimated. The following formula shows the relationship between dissatisfaction and ventilation rate: $PD=359.\exp(-1.83.q^{0.25})$ or $(3.2149 - \frac{lnPD}{1.83})^4 = q$ where PD is predicted dissatisfaction and q is ventilation rate.

References

[1] EPA.(2003). Indoor Air Quality & Student Performance, 1–8

[2] Abbaszadeh, S., Zagreus, L., Lehrer, D., & Huizenga, C. (2006). Occupant Satisfaction with Indoor Environmental Quality in Green Buildings. *Healthy Buildings*, *3*, 365–370

[3] Choi, S., Guerin, D. a., Kim, H.-Y., Brigham, J. K., & Bauer, T. (2014). Indoor Environmental Quality of Classrooms and Student Outcomes: A Path Analysis Approach. *Journal of Learning Spaces*, *2*(2)

[4] World Health Organization

[5]-McNall, P. E., and R. G. Nevins. "Comfort and academic achievement in an airconditioned junior high school: A summary evaluation of the Pinellas County experiment." Ashrae Transactions 73.2 (1967): 3-1

[6] Schoer, L., and J. Shaffran. "A combined evaluation of three separate research projects on the effects of thermal environment on learning and performance." ASHRAE Transactions 79.1 (1973): 97-108\

[7] Gene, L., and Lawrence W. Green. "Measurement and Evaluation Methods for Worksite Stress Management Programs." Stress Management in Work Settings (1987): 109

[8] McIntyre 1980. Indoor climate, Applied Science Publishers Ltd, London

[9] Fanger, P. Ole. "Introduction of the olf and the decipol units to quantify air pollution perceived by humans indoors and outdoors." Energy and buildings 12.1 (1988): 1-6

[10] Clements-Croome, Derek J. "Work performance, productivity and indoor air." SJWEH Supplements 4 (2008): 69-78

[11] Fisk, William J., and Arthur H. Rosenfeld. "Estimates of improved productivity and health from better indoor environments." Indoor air 7.3 (1997): 158-172.

[12] Seppänen, Olli, William J. Fisk, and Q. H. Lei. "Ventilation and performance in office work." Indoor air 16.1 (2006): 28-36

[13] Amérigo, María, and Juan Ignacio Aragones. "A theoretical and methodological approach to the study of residential satisfaction." Journal of environmental psychology 17.1 (1997): 47-57

[14] Humphreys, Michael A. "Quantifying occupant comfort: are combined indices of the indoor environment practicable?." Building Research & Information 33.4 (2005): 317-325

[15] Frontczak, Monika, and Pawel Wargocki. "Literature survey on how different factors influence human comfort in indoor environments." Building and Environment 46.4 (2011): 922-937

[16] Lai, Joseph HK, and Francis WH Yik. "Perception of importance and performance of the indoor environmental quality of high-rise residential buildings." Building and Environment 44.2 (2009): 352-360

[17] Steemers, Koen, and Shweta Manchanda. "Energy efficient design and occupant well-being: Case studies in the UK and India." Building and Environment 45.2 (2010): 270-278

[18] Zalejska-Jonsson, Agnieszka, and Mats Wilhelmsson. "Impact of perceived indoor environment quality on overall satisfaction in Swedish dwellings." Building and Environment 63 (2013): 134-144
[19] Kosonen, R., and F. Tan. "The effect of perceived indoor air quality on productivity loss." Energy and Buildings 36.10 (2004): 981-986.

IOP Conf. Series: Materials Science and Engineering **291** (2017) 012013 doi:10.1088/1757-899X/291/1/012013

Refuge behaviour from outdoor thermal environmental stress and seasonal differences of thermal sense in tropical urban climate

Y Kurazumi[1], J Ishii[2], K Fukagawa[3], E Kondo[4] and A Aruninta[5]

[1] School of Life Studies, Sugiyama Jogakuen University, Chikusa-ku, Nagoya, Aichi 464-8662, Japan
[2] Department of Architecture, Meijo University, Tempaku-ku, Nagoya, Aichi 468-8502, Japan
[3] Department of Architecture, Kyushu Sangyo University, Higashi-ku, Fukuoka, Fukuoka 813-8503, Japan
[4] Department of Architecture, Ariake National College of Technology, Omuta, Fukuoka 836-0097, Japan
[5] Department of Landscape Architecture, Chulalongkorn University, Phyathai Road, Phthumwan, BKK 10330, Bangkok, Thailand

E-mail: kurazumi@sugiyama-u.ac.jp

Abstract. Thermal sensation affects body temperature regulation. As a starting point for behavioral body temperature regulation taken to improve from a poor thermal environment to a more pleasant environment, thermal sense of thermal environment stimulus is important. The poupose of this sutudy is to use the outdoor thermal environment evaluation index ETFe to quantify effects on thermal sensations of the human body of a tropical region climate with small annual temperature differences, and to examine seasonal differences in thermal sensation. It was found temperature preferences were lower in the winter season than in the dry season, and that a tolerance for higher temperatures in the dry season than in the winter season. It was found effects of seasonal differences of the thermal environment appear in quantitative changes in thermal sensations. It was found that effects of seasonal differences of the thermal environment do not greatly affect quantitative changes in thermal comfort.

1. Introduction

The Paris Agreement, a new international legal framework for greenhouse gas reduction, was adopted in December 2015 at the 21st Conference of Parties to the Framework Convention on Climate Change (COP 21) [1]. In order to reduce the serious effects due to climate change, countries throughout the world are aiming for global consensus to keep the average temperature rise to less than 2°C compared to before the industrial revolution. This agreement also mentioned that holding the average temperature rise to less than 1.5°C greatly contributes to reducing the risk.

This situation is affected by the Fifth Assessment Report of the Intergovernmental Panel on Climate Change (IPCC) in April 2014 [2]. If no actions are taken to reduce greenhouse gas emissions, the average temperature of the earth is likely to rise by 3.7°C to 4.8°C compared to before the Industrial Revolution, indicating the possibility of poor outdoor thermal environments. Areas with

Content from this work may be used under the terms of the Creative Commons Attribution 3.0 licence. Any further distribution of this work must maintain attribution to the author(s) and the title of the work, journal citation and DOI.
Published under licence by IOP Publishing Ltd

moderate thermal environments would transform into tropical areas, and there is a possibility that those areas will have smaller annual temperature differences. That is, changes in the four seasons could become less dramatic, with smaller seasonal changes. Physiological functions of the human body are affected by annual temperature differences and nutrition intake ratios, which climate change also affects. In particular, seasonal variation of basal metabolism has been found [3].

Thermal sensation affects body temperature regulation. As a starting point for behavioral body temperature regulation taken to improve from a poor thermal environment to a more pleasant environment, thermal sense of thermal environment stimulus is important. Decrease in the autonomous body temperature regulation function leads to changes in deep temperature. Delay in environmental shelter behavior occurs when there is delay or dullness in thermal sensation reaction to heat and the body's overheating. In order to prevent health damage, it is indispensable to clarify effects of body temperature control load on the human body.

In outdoor environments that are in an unsteady state, thermal adaptation results in a thermal sensation different from the indoor environment. In winter, people prefer a lower temperature than in summer, and tolerate higher temperatures in summer than in winter [4-11]. As in surveys of indoor environments, it is inferred that effects of clothing and environmental history appear as seasonal differences. In addition, in these studies, thermal comfort was targeted to a non-specific overall comfort feeling rather than a thermally unique pleasant sensation. Therefore, it was considered to be thermal comfort, but it cannot be denied that it was a comprehensive comfortable feeling including not only the thermal comfort but also effects of that space. Kurazumi et al. [12, 13] quantified the human body effects by the outdoor thermal environment evaluation index ETFe [14] based on heat balance of the human body. Even actual measurements in outdoor environments limited to thermal sensation showed that people prefer a lower temperature in summer than in winter, and tolerate higher temperatures in summer than in winter.

As mentioned above, physiological functions of the human body are affected by annual temperature differences and the nutrition intake ratio, which climate change also affects. Effects of climate acclimatization due to seasonal variation may appear as seasonal differences in thermal sensation. It is suggested that thermal sensations change due to effects such as acclimatization to thermal environment, but it is not limited to thermal sense according to specific sensation dependent heat exchange. In this study, the pourpose was to compare seasonal differences in thermal sensations by using ETFe to quantify effects on thermal sensations of the human body in a tropical region climate with small annual temperature differences.

Considering that the average temperature of the earth is forecast to rise, studying effects on the human body from outdoor thermal environments in tropical regions may be important for considering how to spend time outdoors in the future. In this study, the effect on thermal sense of the human body were measured in an outdoor thermal environment in winter in Bangkok, Thailand, a tropical region. Then, by comparing it with the results of the study of the dry season in Thailand by Kurazumi et al. [15], it was clarified seasonal differences in effects on the human body.

It is rare to perform a subjective experiment that measures physiological and psychological quantities according to the response of the human body using a large number of subjects. It is difficult to perform an experiment that is hypothesized on a statistical population. Accordingly, as only a small number of subjects was used in this study, the new data are a significant addition to the literature.

2. Experimental design

In order to maintain consistency with the actual measurements in the outdoor environment that Kurazumi et al. [15] performed in the dry season in Bangkok, Thailand, the measurement items were the same as the dry season measurement.

The measurements were carried out in December in the Thai winter season. Mobile measurements were carried out on the campus of Chulalongkorn University, Bangkok, Thailand. A speed of movement slower than a normal walking speed of around 0.7 m/s was used to transport the instruments for measuring the thermal environment. The observation points were drawn at random and the routes to the points were not decided in advance.

After arriving at each measurement point, the subjects waited in a standing posture for five minutes while the test staff setup the measurement instruments for the thermal environment, and preparations for measurement were concluded. Thereafter, the subjects were exposed to the thermal environment in a standing posture for ten minutes. As the subject of the research was the environment surrounding the observation stations, the point of gaze of the subjects was free and unfixed. After ten minutes' exposure, the subjects reported the average thermal sensation and the average thermal comfort of the whole body that they experienced while exposed at the observation point.

For the measurements, observation points were selected with consideration of the condition of the ground surface such as bare ground where the surface is gravel or soil, paved ground such as concrete, asphalt, or blocks, green areas covered with plants, and water surfaces comprising the solid angle of the total celestial sphere (hereafter, green cover ratio), and with consideration of the condition of the sky factor due to buildings, trees, and so on. Five observation points were chosen.

The subjects were 15 healthy young Thais who were 4 young male and 11 young female university students. They were considered to be subjects by most physical standard. In accordance with the Declaration of Helsinki [16], the details of the experiment were explained sufficiently well in advance to the subjects and their consent was obtained for their voluntary participation.

The ETFe [14] is an outdoor thermal environment evaluation index based on the heat balance of the human body. Figure 1 shows the relationship between the ETFe and the main parameters. Accordingly, the calculation of the mean skin temperature used for the calculation of the heat balance of the human body was performed using a weighting coefficient that takes into account the convective

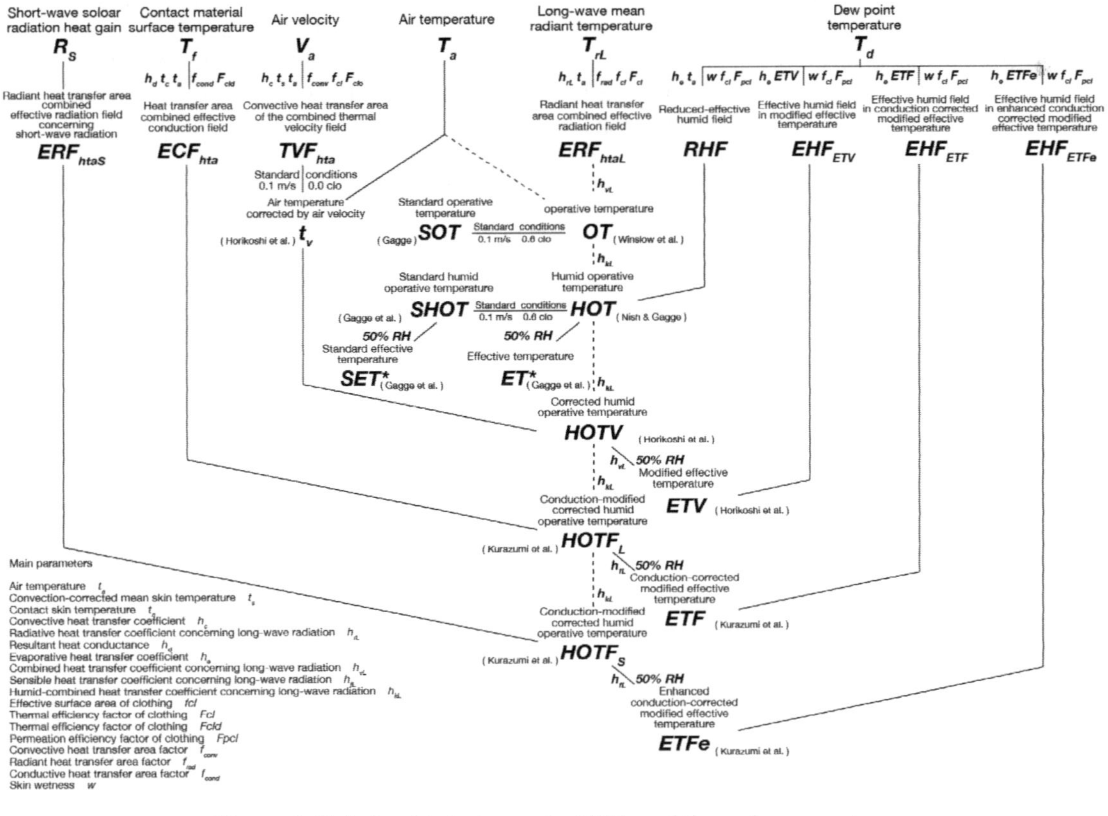

Figure 1. Relationship between the ETFe and the main parameters.

heat transfer area [17]. In calculating the body surface area of the human body, we used the calculation formula of Kurazumi et al. [18]. Then, the calculation of the mean skin temperature used for the physiological response of the human body was performed using a weighting coefficient that takes into account heat conduction [19]. The values of Kurazumi et al. [20] were used for the convective heat transfer area factor, the radiant heat transfer area factor, and the conduction heat transfer area factor for the human body. The value of Miyamoto et al. [21] was used for the projection area ratio of the human body. The values of Kuwabara et al. [22] were used for the radiant heat transfer coefficient and convective heat transfer coefficient of the human body. The value of Hendler et al. [23], found from the reflectance by the skin of electromagnetic waves with wavelengths of 3 μm or more, was used for the emissivity of the human body. The values of Hendler et al. [23] and Elam et al. [24], found from the reflectance by skin of electromagnetic waves with wavelengths of 3 μm or less, were used for the solar radiation absorption coefficient of the human body. The short-wave solar radiation heat gain of the human body is affected by the short-wave solar radiation absorption. According to VDI3787-2 [25], the direct solar radiation absorption of a clothed body is 0.7. However, Watanabe et al. [26] showed that the direct solar radiation absorption of a body clothed in black is 0.76 and that of a body clothed in white is 0.38. Also, the direct solar radiation absorption of other clothing combinations or everyday clothing falls within the range of direct solar radiation absorption for a body in black or white clothing. In this study, a value for the short-wave solar radiation absorption of 0.7, which is standard for a naked body, was used. It was difficult to measure the skin wetness. Therefore, the calculation method used to arrive at these values of skin wetness was the thermoregulation model of Kurazumi et al. [27]. In addition, the value calculated by means of the thermoregulation model of Kurazumi et al. [27] was also used for the physiological quantity that became a missing value. The standard air velocity was considered to be 0.1 m/s and the standard clothing was considered to be a naked body (0clo) in this research. The ETFe index theoretically proposed by Kurazumi et al. [14, 28] was calculated from weather observation values, the skin temperature of the human body, and the clothing value. It was difficult to measure the skin wetness. Therefore, the calculation method used to arrive at these values of the skin wetness was the thermoregulation model of Kurazumi et al. [29].

These explanatory variables of thermal sensation are direct environmental stimuli to thermal sensation, and greatly affect physiological and psychological reactions of the human body: temperature, humidity, wind speed, short wavelength solar radiation heat, long wavelength radiant heat, surface temperatures of materials that contact the human body. In this study, considering that the psychological reaction of the human body in an outdoor environment is likely to cause many disturbances and variations, as a criterion for comparing explanatory variables for deriving results of regression analysis with greater usefulness from practical point of view, the significance probability was set to 10%. JMP 12.2.0 (SAS Institute Japan) was used for statistical analysis.

3. Results and discussion

3.1. Observed thermal environment summary

The temperature dropped under 30°C at some observation points. The climate of Bangkok was in winter, during which temperatures were relatively low. There was a large amount of downward short-wave solar radiation of open observation points with a high sky percentage, but there was much less downward short wavelength solar radiation at observation points where sunlight was obstructed by trees etc. with low sky ratio. The effect of sunlight is very noticeable.

There were great amounts of long-wavelength radiation at all the observation points. Even if one is sheltered in a place with solar radiation shielding avoiding short wavelength solar radiation, one still receives much heat radiation. At observation points with concrete pavement where there is high heat capacity and strong effects of short wavelength solar radiation, the ground surface temperature in the vicinity of the observation point is remarkably higher than other observation points. There is very little contact area between the standing posture human body and the ground surface, but it is conjectured that the heat acquisition of the human body by heat conduction strongly affects contact skin temperature. In addition, it is conjectured that the high surface temperature in the vicinity of the human body strongly affects the heat radiation amount because the shape factor increases. The trend of the measured results of short wavelength solar radiation and long wavelength radiation at the observation points was similar to the dry season results of Kurazumi et al. [15].

The measured average value of relative humidity is 60.4%, the standard deviation is 7.4%, and no notable difference in actual measurement results is shown. Therefore, there may be weak effects on the sensed temperature by the difference in evaporative heat exchange amount.

The average wind speed was a relatively calm 2m/s or less through all the observations. The standard deviation was 0.4 m/s. Therefore, difference in convective heat exchange amount may weakly affect sensed temperature. This may be affecting thermal sensation as a factor of environmental change.

The following discussion will clarify seasonal differences of effects on the human body, by comparison with results of research in Thailand's dry season by Kurazumi et al. [15].

3.2. Relationship between ETFe and thermal sensation

Figure 2 shows the relationship between ETFe and Thermal sensation. It shows a tendency for people to report on the "hot" side when ETFe becomes higher. Thermal sensation shows a wide distribution from "cool" to "hot", showing remarkable individual differences.

When ETFe is in the range 30-40°C, thermal sensation in the range from "Slightly cool" to "Slightly warm" tends to be reported. Similar to what Kurazumi et al. [12, 13, 28] points out, it seems that expectation for comfort in the outdoor space is originally low, judging that it is not a comfortable thermal environment, so ETFe was acceptable even under high thermal environmental conditions. Although it is uncomfortable thermally, it could be that people tolerate a wider range of thermal environments in outdoor spaces where they can change the selection of place and viewpoint by their own will. That is, it is thought that due to the degree of freedom of behavior in the experimental environment, large fluctuations in the psychological reactions of the human body occurred in the

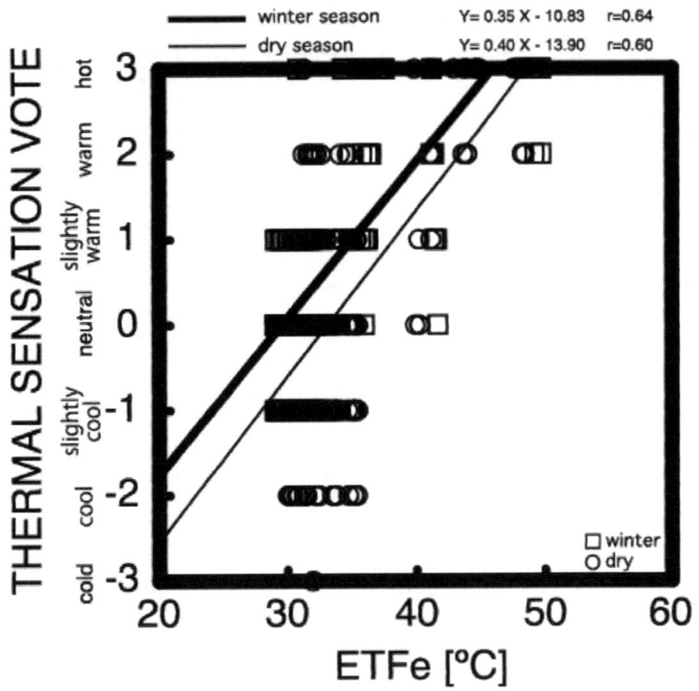

Figure 2. Relation between ETFe and thermal sensation.

outdoor space [13]. The outdoor thermal environmental factors of short wavelength solar radiation, long wavelength radiation and air current are uneven and non-targeted evaluation factors. These evaluation factors may contribute to local effects on the human body, even if heat balance in the whole body is the same.

Focusing on the regression line, there is no large difference in the slope of the regression line, and it seems there is a tendency that in winter, people prefer a thermal environment that is about 4°C lower than in the dry season. People like winter temperatures that are lower than they like in summer, and tolerate higher temperatures in summer than in winter, as found by Kurazumi et al. [12, 13]. The climate of Bangkok in Thailand is hottest season in the dry season, so it can be regarded as summer. Therefore, this study also inferred that effects of environmental history appear as seasonal differences.

The result of testing parallelism of the regression line was p>0.10, and no significant difference in parallelism of the regression line was shown. The result of testing homogeneity of regression was p<0.10, indicating a significant difference in homogeneity of the regression line. Therefore, in the range of the results of ETFe in this study, it seems that effects of seasonal differences, which seem to be lower in winter than in the dry season, appeared in the amount of change to thermal sensation.

3.3. Relationship between ETFe and thermal comfort

Figure 3 shows the relationship between ETFe and thermal comfort. As ETFe increases, the strength of "uncomfortable" tends to increase. The feeling of comfort shows a wide distribution from "comfortable" to "extremely uncomfortable", showing remarkable individual differences.

In this research, an experimental method that isolates environmental history as much as possible was used, but it seems that there are effects of thermal environment history, and an expectation of environmental sheltering thermal environment accompanying mobile observation. Kurazumi et al. [12] states that 38.5°C is the upper limit ETFe of the optimal environmental range in summer. After ETFe exceeds that upper limit ETFe temperature, the variation becomes large. Therefore, subjects subjects were instructed to report thermal comfort according to specific sensation dependent heat exchange, but

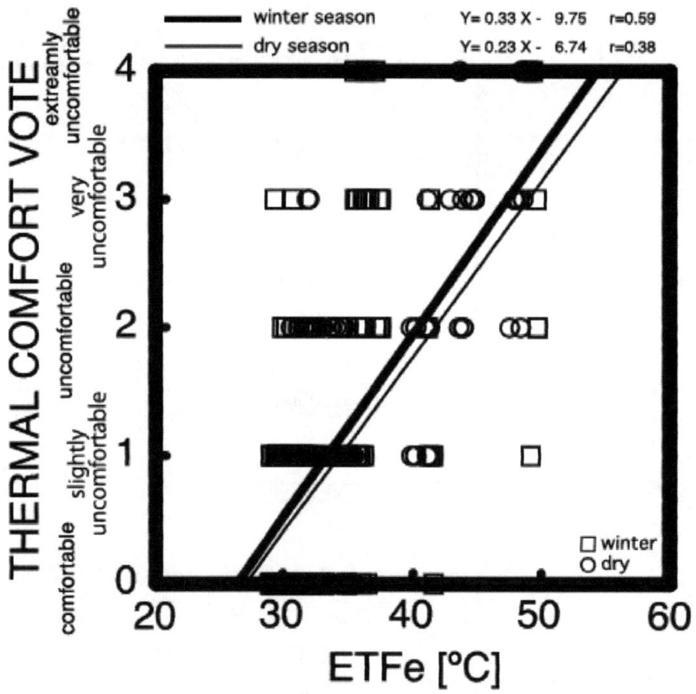

Figure 3. Relation between ETFe and thermal comfort.

the possibility that they are reporting comprehensive thermal comfort including the effects of spatial impression cannot deny. As pointed out by Kurazumi et al. [13], outdoor thermal environmental factors (short-wavelength solar radiation, long-wavelength radiation, and airflow) are uneven and non-targeted evaluation factors, and effects of parameters of these environmental factors could cause variations in human body reactions. It is possible for the human body to experience discomfort due to variation in environmental factors, and this experience can be considered to induce relative comfort. Accordingly, variations with an effect that mitigates thermal action are considered to be variables for discomfort with regard to other environmental factors as well.

Focusing on the regression line, similar to what Kurazumi et al. [12, 13, 28] points out, it seems that expectation for comfort in outdoor spaces is originally low, and people judge that it is not a comfortable thermal environment, so ETFe is tolerable even under high thermal environmental conditions. The slope of the regression line is slightly greater in winter than in the dry season, so it seems that reactions to thermal sensation for thermal comfort appears more strongly in winter than in the dry season, even at almost the same ETFe.

The result of testing parallelism of the regression line was $p>0.10$, and no significant difference was shown in parallelism of the regression line. The result of testing homogeneity of regression was $p>0.10$, showing no significant difference in homogeneity of the regression line. Therefore, it seems that within the range of ETFe results of this experiment, effects of seasonal differences do not exert a large influence on the amount of change to thermal comfort.

The laboratory experiment of Fanger [30] found no seasonal differences in thermal comfort. However, according to a questionnaire survey at the office of Miura et al. [31], the thermal neutral temperature has a seasonal difference, which is higher in summer than in winter. In subject experiments in outdoor environments of Kurazumi et al. [12, 13], in Japan where the annual difference in temperature is larger than in Bangkok, people tolerate higher temperatures in summer than in winter, and they state that thermal comfort has seasonal differences. In 2016, the annual difference in temperature in Bangkok was around 4°C, but the annual difference in temperature in Tokyo was around 21°C [21].

4. Conclusion

Effects on thermal sensation of the human body from the outdoor thermal environment of a tropical region with small annual temperature differences was quantified by the total thermal environment evaluation index ETFe of the outdoor environment, and seasonal differences of thermal sensation were compared. Thermal sensation affects body temperature regulation. As a starting point for behavioral body temperature regulation taken to improve from a poor thermal environment to a more pleasant environment, thermal sense of thermal environment stimulus is important. The findings obtained in this study are shown below.

In the dry season. People prefer lower temperatures in winter than in the dry season, and tolerate higher temperatures in the dry season than in winter. It seems that effects of seasonal differences appeared in the amount of change to thermal sensation.

It is possible for the human body to experience discomfort due to variation in environmental factors, and this experience can be considered to induce relative comfort. It is thought that one cannot rule out the possibility that people are reporting comprehensive thermal comfort including effects of space impression. on thermal comfort according to specific sensation dependent heat exchange. In other words, it seems that effects of seasonal differences are not likely to greatly affect the thermal sensation change amount for thermal comfort.

When planning the safety urban development, the consideration for the outdoor environment becomes an important matter. Thus, it is to clarify the psychological and the physiological reactions of the human body in outdoor thermal environment. This study inferred that effects of environmental history appear as seasonal differences. There is a need for further study that takes seasonal acclimatization into account is necessary.

Acknowledgments

We would like to express our sincerest gratitude to the study subjects who participated in the present study.

References

[1] UnitedNations.FCCC/CP/2015/L.9/Rev.1:http://unfccc.int/resource/docs/2015/cop21/eng/l09r01.pdf
[2] IPCC Working Group I Contribution to AR5. Summary for Policymakers of Climate Change 2013: The Physical Science Basis: http://www.climatechange2013.org/images/report/WG1AR5_SPM_FINAL.pdf
[3] Shimaoka A, Machida K, Kumae T, Sugawara K, Kurakake S, Okamura N and Suemune J 1987 Seasonal variation of basal metabolism Japanese Journal of Biometeorology 24 (1) pp 3-8
[4] Höppe P 2002 Different aspects of assessing indoor and out- door thermal comfort Energy and Buildings 34 pp 661-665
[5] Nikolopoulou M and Steemers K 2003 Thermal comfort and psychological adaptation as a guide for designing urban spaces Energy and Buildings 35(1) pp 95-101
[6] Spagnolo J and de Dear R J 2003 A field study of thermal comfort in outdoor and semi-outdoor environments in subtropical Sydney Australia Building and Environment 38(5) pp 721-738
[7] Nakano J and Tanabe S 2004 Thermal comfort and adaptation in semi-outdoor environments ASHRAE Transactions 110(2) pp 543-553
[8] Ahmed K S 2003 Comfort in urban spaces: defining the boundaries of outdoor thermal comfort for the tropical urban environments Energy and Buildings 35(1) pp 103-110
[9] Oliveira S and Andrade H 2007 An initial assessment of the bioclimatic comfort in an outdoor public space in Lisbon International Journal of Biometeorology 52(1) pp 69-84
[10] Lin T P and Matzarakis A 2008 Tourism climate and thermal comfort in Sun Moon lake, Taiwan International Journal of Biometeorology 52(4) pp 281-290
[11] Cheng M, Lo J and Chen S 2011 Investigation of the effect of climatic adaptation on users' thermal comfort requirement in outdoor space International Journal of the Physical Sciences 6(25) pp 6042-6052
[12] Kurazumi Y, Tsuchikawa T, Kondo E, Ishii J, Fukagawa K, Yamato Y, Tobita K, Ando Y, Matsubara N and Horikoshi T 2012 Thermal comfort zone in outdoor environment Journal of Human and Living Environment 19(2) pp 115-127
[13] Kurazumi Y, Ishii J, Kondo E, Fukagawa K, Bolashikov Z D, Sakoi T, Tsuchikawa T, Matsubara N and Horikoshi T 2014 The influence of outdoor thermal environment on young Japanese female International Journal of Biometeorology 58(5) pp 963-974
[14] Kurazumi Y, Fukagawa K, Yamato Y, Tobita K, Kondo E, Tsuchikawa T, Horikoshi T and Matsubara N 2011 Enhanced conduction-corrected modified effective temperature as the outdoor thermal environment evaluation index upon the human body Building and Environment 46(1) pp 12-21
[15] Kurazumi Y, Ishii J, Fukagawa K and Aruninta A 2015 The influence of tropical urban climate upon the human body International Joint-Conference of SENVAR-iNTA-AVAN 2015 (SIA 2015) pp 2-105-2-114
[16] World Medical Association. WMA Declaration of Helsinki - Ethical Principles for Medical Research Involving Human Subjects: http://www.wma.net/en/30publications/10policies/b3/index.html
[17] Kurazumi, Y, Tsuchikawa, T, Torii, T, Kakutani, K, Matsubara, N and Horikoshi, T 2004 Weighting coefficients for calculating mean skin temperature when considering convective heat transfer areas Journal of the Human-Environmental System 7(1) pp 19-28
[18] Kurazumi, Y, Horikoshi, T, Tsuchikawa, T and Matsubara, N 1994 The body surface area of Japanese Japanese Journal of Biometeorology 31(1) pp 5-29
[19] Kurazumi Y, Matsubara N, Furukawa N, Fujiwara M, Ue A, Ueki Y, Nagai H and Yamamoto S

1998 Japanese Weighting coefficients for calculating mean skin temperature in relation to posture Japanese Journal of Biometeorology 35(4) pp 121-132
[20] Kurazumi Y, Tsuchikawa T, Matsubara N and Horikoshi T 2008 Effect of posture on the heat transfer areas of the human body Building and Environment 43(10) pp 1555-1565
[21] Miyamoto S, Horikoshi T and Hirokawa Y 1998 Projected area factors of the human body at standing posture under different clothing conditions Journal of Architecture, Planning and Environmental Engineering (Transactions of AIJ) 513 pp 47-52
[22] Kuwabara K, Mochida T, Kondo M and Matsunaga K 2001 Measurement of man's convective heat transfer coefficient by using a thermal manikin in the middle wind velocity region Journal of Human and Living Environment 8(1, 2) pp 27-32
[23] Hendler E, Crosbie R and Hardy J D 1958 Measurement of heating of the skin during exposure to infrared radiation Journal of Applied Physiology 12 pp 177-185
[24] Elam R, Goodwin D W and Lloyd W K 1963 Optical properties of the human epidermis Nature 198(4884) pp 1001-1002
[25] VDI 2008 VDI 3787-2, Environmental meteorology – Methods for the human biometeorological evaluation of climate and air quality for urban and regional planning a regional level – Part 1: Climate Beuth Berlin Germany
[26] Watanabe S, Horikoshi T and Tomita A 2010 Measurement of solar radiation absorptamce of clothed human body in outdoor Japanese Journal of Biometeorology 47(4) pp 165-173
[27] Kurazumi Y, Sakoi T, Tsuchikawa T, Fukagawa K, Bolashikov Z D and Horikoshi T 2014 Behavioral thermoregulation model for evaluation of outdoor thermal environment Journal of Ergonomics 4(1) pp 1-14
[28] Kurazumi Y, Tsuchikawa T, Matsubara N, Kondo and Horikoshi T 2011 Evaluation of enhanced conduction-corrected modified effective temperature ETFe as the outdoor thermal environment evaluation index Energy and Buildings 43(10) pp 2925-2937
[29] Kurazumi Y, Sakoi T, Tsuchikawa T, Fukagawa K, Bolashikov Z D and Horikoshi T 2014 Behavioral thermoregulation model for evaluation of outdoor thermal environment Journal of Ergonomics 4(1) pp 1-14
[30] Fanger P O 1970 Thermal comfort Danish Technical Press Copenhagen Denmark
[31] Miura T 1968 Studies on the optimum temperature, - On some factors affecting the shift of optimum temperature - The Journal of science of labour 44(8) pp 431-453
[32] Japan Meteorological Agency. ClimatView:
http://www.data.jma.go.jp/gmd/cpd/monitor/climatview/frame.php

A Review of Criteria for Outdoor Classroom in Selected Tertiary Educational Institutions in Kuala Lumpur

Maheran Y[1], **Fadzidah A**[2], **Nur Fadhilah R**[3] and **Farha S**[3]
[1]Asst. Professor, Department of Landscape Architecture, IIUM, Kuala Lumpur, Malaysia
[2] Asst. Professor, Department of Architecture, IIUM, Kuala Lumpur, Malaysia
[3]Postgraduate Student, Department of Landscape Architecture, IIUM, Kuala Lumpur, Malaysia
[3]Postgraduate Student, Department of Landscape Architecture, IIUM, Kuala Lumpur, Malaysia
E-mail: maheran@iium.edu.my[1], fadzidah@iium.edu.my[2], fadhilahrozali@gmail.com[3], farha.rha9@gmail.com[3]

Abstract. A proper design outdoor environment in higher institutions contributes to the students' learning performances and produce better learning outcomes. Campus surrounding has the potential to provide an informal outdoor learning environment, especially when it has the existing physical element, like open spaces and natural features, that may support the learning process. However, scholarly discourses on environmental aspects in tertiary education have minimal environmental inputs to fulfill students' needs for outdoor exposure. Universities have always emphasized on traditional instructional methods in classroom settings, without concerning the importance of outdoor classroom towards students' learning needs. Moreover, the inconvenience and discomfort outdoor surrounding in campus environment offers a minimal opportunity for students to study outside the classroom, and students eventually do not favor to utilize the spaces because no learning facility is provided. Hence, the objective of this study is to identify the appropriate criteria of outdoor areas that could be converted to be outdoor classrooms in tertiary institutions. This paper presents a review of scholars' work in regards to the characteristics of the outdoor classrooms that could be designed as part of contemporary effective learning space, for the development of students' learning performances. The information gathered from this study will become useful knowledge in promoting effective outdoor classroom and create successful outdoor learning space in landscape campus design. It I hoped that the finding of this study could provide guidelines on how outdoor classrooms should be designed to improve students' academic achievement.

Keywords: outdoor classroom, campus landscape design, informal learning environment, teaching and learning

1. Introduction

An outdoor classroom is characterized an outdoor education facility that encourages teachers, learners, and educators to gain better knowledge and provide alternatives to learn and understand the interconnectedness of the natural resources [1]. In higher institutions, this approach is described as a tool that allows students and lecturers move their educational curriculum outdoors, and experience the natural and physical-created elements in natural settings.

Content from this work may be used under the terms of the Creative Commons Attribution 3.0 licence. Any further distribution of this work must maintain attribution to the author(s) and the title of the work, journal citation and DOI.
Published under licence by IOP Publishing Ltd

As mentioned in [2], stated that the outdoor classroom provides opportunities to experience 'natural studies environment', as the physical environment creates positive impact towards student's learning possibilities and learning outcomes. Moreover, good landscape campus setting is essential to promote the physical and cognitive development of students and improve learning abilities.

The previous setting of Malaysian university campus settings is designed to accommodate formal teaching studies, like lecture halls, air-conditioned classrooms, and laboratories, with least consideration on integration with campus ground [2]. Studies by [4] also agreed that there is a lack of natural settings from primary schools to university level. Inconsiderate outdoor area and pocket spaces for a student's learning led to distinct interaction between the quality of the outdoor environment and the quality of life in the campus [3]. Hence, students tend to spend their break time indoor rather than outdoor, as the outdoor spaces are not functionally designed to support their learning activities.

This research has the objective to identify the criteria of the outdoor classroom in selected tertiary institutions, that is, the identification of the significant element of the outdoor classroom that could achieve physical and psychological realms of the students. The statement supported by scholars [19][10][14] underline physical element of outdoor classroom that contributes not only students experience, but also enhance educational pedagogy and students behavior settings.

2. Methods

Table research has been carried out to gather secondary data for this research, focusing on the significance criteria of the outdoor classroom in tertiary institutions in Kuala Lumpur. The aim of the analysis is to expand the understanding of the criteria of the outdoor classroom, and their relationship with student's learning activities. Review of articles was done to analyze and categorize the key parameters and significant attributes of the outdoor classroom. They are presented in table format, and selected were further elaborated in the findings section.

3. Outdoor Classroom in Tertiary Institutions

Malaysian campus settings are still applying traditional education practices which disservice to the academic aspirations of 21st century teaching institutions; that is, enriches students' lives with exposure to the natural environment and emphasizes the actual experience gain outside of the classroom [5]. The setting of outdoor learning in the university is important to encourage students to utilize the spaces vigorously and also to meet student's changing study needs [2]. Thus, by means, effective outdoor spaces that provides varying learning activities and promotes technological accessibility to aid students learning processes.

According to [4][6], connecting students with outdoor classroom helps to build a sense of belonging, develop social connections among peers and enhance personal skills performances including positive self- concept. The studies also highlight the advantages of the outdoor classroom in the learning environment which is (i) broadening educational performances (ii) developing individual potential (iii) motivating towards self-directed learning and also (iv) sharpening students' intellectual ability and psychological development.

Further view, studies by [19] acknowledges the component of outdoor spaces for active learning is according to the experiences and understanding of the connectivity between the architecture and landscape environment. Table 1 illustrates the three key features on an outdoor classroom for student – centered approaches.

Table 1. Key features of design spaces

Key features of design spaces for active learning
1. **Sense of belonging**
-The space that accommodates diverse students and public
-Maximize the use of day lighting to reduce energy use
-Multipurpose spaces for varying learning activities
-Comfortable spaces build a feeling of connection towards the environment
2. **Flexibility and multi-use spaces**
-Movable furniture to incorporate multifunction activities
-Highly flexible, self-contained and free distraction spaces
-Increasing flexibility spaces for specific functions / multifunction activ ities
-Design features that maximize user control
3. **The uses of non-classroom spaces for learning**
-Design spaces to overcome functions and services
-Space availability merges social interaction
-Usable transaction space between indoor and outdoor

The key features outlined 3 components of the establishment of learning spaces. The features are essential in designing the outdoor spaces with the building settings to encourage connections between campus layout and physical environment, intellectual culture and effective learning environment [2]. Therefore, a good campus environment does not only promotes positive learning attitudes, but also enhance students learning performances as preparations for students to overcome university challenges and beyond. In order to achieve the student – centered learning practices, the criteria of the outdoor classroom seem relevant to adequate the approach by considering the needs of students and educators, making it possible to support the transformation of the learning environment.

3.1 Criteria of Outdoor Classroom

A total of ten research articles was reviewed and included in the analysis. Table 2 shows the characteristics of the outdoor classroom in campus institutions, and the summaries of the literature reviewed are presented in the form of matrix data in Figure 1.

According to the analysis, most of the scholars agree that multi uses of outdoor classroom provide varied activities for students to achieve their physical and psychological development. Followed by the comfortable outdoor spaces and accessibility as most essential elements in the outdoor classroom. The comfortable spaces with the use of natural lighting controlled temperature and supported with suitable furniture enable students to utilize the spaces for any occasions[7]. Accessibility in outdoor spaces offers good connectivity movement in campus area for students in daily basis needs[3]. The continuity, directness and multiple choices of movement contribute to the increase of students walking in the campus and avoid massive vehicle usage.

Further view, the elements of calm, variety furniture, outdoor design element and technology and ICT tool changes learning spaces into flexible outdoor learning spaces. These findings support the previous study by [8] indicates that students much prefer outdoor spaces with movable furniture and the ability to use technology tools freely for learning activities. Besides, the calm environment allows students experience mental stability, reduce stress and relaxed mind [9]. Lastly, the criteria of outdoor classrooms, highlight the aesthetic element and social relation that affect student social integration at university campuses. As mentioned in [10] that describe the importance of social relation including students' interaction with peers and lecturers and the involvement in extra-curricular activities. These findings also support previous studies by [8] indicates that outdoor spaces act as a community social space; engage students and faculty in participating in outdoor learning and forging connections, hence create a sense of belonging towards them.

Further examined, the physical features in the outdoor classroom enhance students towards learner-centered approach and promote the contemporary design element that suits their learning activities.

Table 2. Characteristic of the Outdoor Classroom

Authors	**Characteristics of outdoor classroom**	
Acker, S., and Miller, M., (2005)	1. Flexibility 2. Comfort 3. Ergonomic seating 4. 'Noise Zone'	5. Pervasive technology 6. Access learning media 7. Accessibility
Hunley, S., and Schaller,M., (2009)	1. **Individual study** -Free from distraction and noise -Comfortable -Pleasing aesthetic -Variety furniture -Good lighting	2. **Group work discussion** -Soft furniture for discussion -Controllable facilities -Spaces integrate needs and desires
Riddle, M., and Soulter, K., (2012)	1. **Comfort** spaces 2. **Aesthetical** values-simplicity and harmony 3. Physical and psychological flow of **movement**	4. **Equity**- the space support PWDs movement 5. **Blending** of technology tools and facilities 6. **Repurposing**-accommodate different activities as possible
Ibrahim, N., Fadzil, N., & Saruwono M.,(2013)	1. Environmentally protected spaces 2. Flexibility 3. Furniture for learning	4. ICT tools and facilities 5. Varies design element beyond functional needs
Zandvliet, D., (2013)	1. **Spatial environment** -Space for physical and learning needs 2. **Scale and aesthetics** -Accessibility and attractiveness of space	3. **Ambient factors** -Comfortable designed spaces 4. **Architectural element** -Flow and layout of spaces 5. **Visual element** -Availability and adaptability lighting in outdoor setting
Matloob, F., Sulaiman, A., Ali, T., Shamsuddin, S., & Mardyya, W., (2013)	1. **Ease of movement** -Connectivity -Legibility -Comfort and safety 2. **Sense of identity** -Local style -Native plants	3. **Accessibility** -Distance between spaces -Continuity -Ease finding spaces 4. **Quality of public realm** -Cleanliness -Overall appearance -Quietness
Ibrahim, N., and Fadzil, N., (2013)	1. **Creative and innovative** space 2. **Multipurpose** outdoor area 3. **Integration** between indoor and outdoor	4. Designed space provide activities 5. Outdoor spaces as a **focal point** 6. Shaded spaces 7. **Technology** Accessibility
Sedaghatnia, et. al., (2014)	1. Legibility 2. Social relation	3. Space quality 4. Accessibility 5. Comfort
Ali, S., Rostam, K., & Awang, A., (2014)	1. **Thermal comfort** -reduced temperature surroundings 2. Aesthetic value 3.**Natural elements** -students interact with nature	4. Identity of the space 5. **Educational related concept** -stimulate student's learning interest 6. Safety and security 7. Cleanliness
Hossini, S., Azemati, A., Elyasi, N., & Mozaffar, F., (2014)	1. Accessibility 2. Safety 3. Comfort 4. Comfortable furniture 5. Calm for private and meeting space	6. Eliminating non-emergency preventives 7. Multi-functional spaces 8. Sense of place

Figure 1. Matrix Data of the Characteristic of Outdoor Classroom

Criteria of outdoor classroom		Acker and miller	Hunley and Schaller	Riddle and Soulter	Ibrahim et. al.,	Zandvliet	Matloob et. al.,	Ibrahim and Fadzil	Sedaghatnia et. al.,	Ali et. al.,	Hossini et. al.,
Flexibility & Multiuse of space		●	●	●	●	●		●			●
Comfort	Comfortable	●	●	●		●	●		●		●
	Calm	●	●				●				●
	Variety furniture	●	●		●						●
	Thermal comfort								●		
	Visual element		●			●					
	Cleanliness						●			●	
	Facilities	●	●								
	Security & safety						●			●	●
	Integration		●					●			●
Movement	Accessibility	●				●	●		●		●
	Connectivity		●	●			●				
	Equity			●							
	Legibility						●		●		
	Continuity						●				
Technology & ICT Tools		●		●	●			●			
Aesthetic	Simplicity & harmony		●	●							
	Natural element				●						
	Design element				●	●		●		●	
	Sense of identity						●			●	●
	Focal point							●			
Social Relation									●		

3.2 Design Principles of Outdoor Classroom

In addition, information on the criteria of the outdoor classroom, studies by [18][8][17][16] outline the design principles in the creation of contemporary learn spaces, that offer an inclusive environment and robust basis in assessing alternative concepts of outdoor classroom settings. Table 3 illustrates the design principles for learning spaces.

Studies by [8] elaborate the use of technology brought unique capabilities to learning. Students recognize the technology supports that provide the fastest way of learning, especially in completing the project that needs multiple works. Moreover, students share the information, communicate with peers and access the information through mobile devices and wireless network. The increasing used of technology changes the outdoor spaces to be more flexible and networked, as the learning activities can occur in any place.

Table 3. The Design Principle for Learning Spaces

Author	Design Principles for Learning Spaces
Jamieson et. al., (2005) **(Adoption of multi-disciplinary approaches)**	1. Outdoor spaces for **multiuse** concurrently 2. **Flexibility** designed within each space 3. Design settings that have **varied facilities** 4. Design **features and functions** to maximize teacher and student control 5. **Arrangement** spaces to support student's curricula 6. Learning space design that provides **accessibility** for students
Oblinger, D. (2006) **(Learner-centered approach to the design facilities)**	2. Support **multiple types of learning activities** 3. Enable **connection** within surrounding 4. Accommodate **information technology** 5. Design spaces for **comfort, safety** and **functionality** 6. Reflect institutional **values**
Siddall, S. (2006)	1. **Diversity** spaces for learning activities 2. **Versatile** 3. **Technologically** reliable settings 4. **Well-maintained facilities** 5. The spaces should be **universal** in space and time 6. **Effective** learning space for student's multi activities 7. Sufficient **resources** allocated for learning spaces
Radcliffe et. al., (2009) **(Spaces within an educational building)**	1. **Flexible** outdoor spaces that accommodate current and evolving pedagogies 2. **Future proofed** settings that enable space to reconfigure and adapt external changes 3. The space that **technology** facilitated 4. Outdoor settings that **inspire** and promote **creativity** to users 5. **Enterprising**- Outdoor settings that support different purposes

The multi-use spaces in higher institutions are expanding into numerous space types to create a more attractive environment and more useable space [10]. The study conducted by [5] listed space types that is; (i) spaces for group teaching and learning with varying sizes and layouts, (ii) simulated environment with technological capacity,(iii) highly interactive virtual environments with ICT facilities, (iv) informal learning environment where peer-to-peer conversation, (v) clusters for student group work. (vi) private learning spaces and (vii) areas outside the buildings for individual or small group activity.

The third significant design principles as mentioned above is the flexible spaces that give opportunities to educators and learners choices to have multidisciplinary activities. According to [18][16], flexible spaces allow students to experience both indoor and outdoor learning environments, and transform the space from small group space to a large group discussion space. Thus, the spaces cater student-centered learning pedagogies and could be modified to suit with preferred learning styles and their needs.

4. Conclusion

Today's world concern on refining the educational perspectives embraces learning resources and technology as learning tools from primary to university level. Hence, outdoor classroom gives positive benefits to the students in all ranges of educational levels as the quality of the learning environment is affected the student's learning performances.

This review reveals that there is a considerable research suggesting the interacting parameters between the outdoor classroom and student's learning activities. Overall, the reviews suggest that the outdoor classroom can provide progressive outcomes in physical, psychological and emotional students in tertiary educations. The criteria of the outdoor classroom are very important in the campus landscape design as it is influencing students for learning, educating, meeting, exploring as well relaxing.

5. Acknowledgment

This work was supported in part of the grant as RIGS under RMC, International Islamic University Malaysia (IIUM).

References

[1] Kimbro, C. 2006. Developing an Outdoor Classroom to Provide Education Naturally. The University of Tennessee Extension

[2] Ibrahim, N., Fadzil, N., Saruwono, M. 2013. *Learning Outside Classrooms on Campus Ground: Malaysia*, 97-109

[3] Matloob, F., Sulaiman, A., Ali, T., Shamsuddin, S., Mardyya, W. 2013. *Sustaining Campuses Through Physical Character- The Role of Landsape*. PSYSOC 2013, 282-290

[4] Mirrahimi, S., Tawil, N., Abdullah, N., Surat, M., Usman, I. 2011. *Developing Condusive Sustainable Outdoor Learning: The impact of Natural Environment on Learning, Social and Emotional Intelligence*. The 2nd International Building Control Conference, Procedia Engineering (20) 389-396

[5] Temple, P. 2007. Learning spaces for the 21st century. A review of the literature. Centre for Higher Education Studies, Institute of Education, University of London

[6] Harun, M., Salamuddin, N. 2013. Applying Elements of Outdoor Education in Teacher Education Innovation. Asian Social Science **9** (16). 15-21

[7] Riddle, M., Souter, K. 2012. Designing informal learning spaces using student perspectives. *Journal of Learning Spaces* **1** (2)

[8] Oblinger, D. 2006. Learning Spaces. Educause e-Book www.educause.edu/learningspaces

[9] Ali, S., Rostam, K., Awang, A. 2014. School Landscape Environments in Assisting the Learning Process and in Appreciating the Natural Environment. *Asean – Turkey ASLI.* International Conference on Quality Life,AQoL. 189-198

[10] Sedaghatnia, S., Lamit, H., Abdullah, A., Ghahramanpouri, A. 2014. Experience of Social Inclusion among Students in University Campuses of Malaysia. Asian Conference on Environment- Behavior Studies, AcE-Bs. 89-98

[11] Acker, S., Miller, M. 2005. Campus Learning Spaces: Investing in How Students Learn. EDUCAUSE Center for Applied Research, **2005** 8. 1-11

[12] Hunley, S., Schaller, M., (2009). Assessment the key to Creating Spaces that Promote Learning. EDUCAUSE Review (2009) 26-34

[13] Ibrahim, N., Fadzil, N. 2013. Informal Setting for Learning on Campus: Usage and preference. Asia Pacific International Conference on Environment- Behavior Studies, AicE-Bs. 344-351

[14] Zandvliet, D. 2013. PLACES and SPACES: Case studies in the evaluation of post-secondary,place-based learning envronments. Studies in Educational Evaluation. 18-28

[15] Hossini, S., Azemati, S., Elyasi, N., Mozaffar, F. 2014. The Effect of the Vitality Level of University Campuses on Increasing Social Interactions and Making. Asia Conference on Environment- Behavior Studies, AcE-Bs. 225-233

[16] Jamieson, P., Fisher, K., Gilding, T., Taylor, P., Trevitt, A. 2000. Place and Space in the Design of New Learning Environment. HERDSA 19. 221-237

[17] Siddall, S. 2006. The Denison Learning Space Project, Mission and Guiding Principles http://www.denison.edu/learningspaces/mission.html

[18] Radcliffe, D., Wilson, H., Powell, D., Tibbetts, B. 2009. Learning Spaces in Higher Education: Positive Outcomes by Design. The University of Queensland

[19] Tahir, M., Abdullah, N., Usman, I., Che-Ani, A., Mohd Nor, M., Surat,M. 2009. Constructing Place and Space in the Design Learning Environments For PBL in Malaysian Universities, AJTLHE, 1, 26-34

Supply Chain Management on IBS Implementation in Klang Valley Construction Industry: Challenges and Issues

Mohd Azrizal Fauzi[1], **Sulaiman Hasim**[1], **Anizah Awang**[1], **Ahmad Ruslan Mohd Ridzuan**[1] and **Juzailah Nur Yunus**[1]
[1]Faculty of Civil Engineering, Universiti Teknologi MARA (UiTM), Malaysia
E-Mail: azrizal_fauzi@yahoo.com

Abstract. Industrialized Building System (IBS) is a system where the components of the building are manufactured in a factory and it will be transported to the site to form the structures. The supply chain management (SCM) is a system where the delivery flows of the IBS products from manufacturers to the site. The aim of this research is to identify the major challenges and to analyze the issues on IBS implementation in SCM in Klang Valley from the manufacturers perspective. The methodology used in this paper is based on primary data through questionnaire and interview. Questionnaires were sent to the Manufacturers. It can be concluded that this paper attempts to present more on the challenges and issues that those companies of manufacturers faced during their success journey in finding integration in their supply chain. The main contributions of this paper are integrating all the supply chain integration challenges and issues on IBS. Therefore, these contributions will be helpful for the organization of manufacturers and IBS players that establish the integration in their SCM.

Keywords: Supply Chain Management, Construction Industry, IBS, Manufacturers.

1. Introduction

IBS is a construction system that manufactured in a factory to minimize the additional work at the site [1]. The use of IBS can provide speedier work completion due to the replacing of the components on-site construction. Furthermore, the government of Malaysia has started the implementation of IBS in their construction contract for IBS towards 2010. Therefore, Malaysia can improve and experience more in the construction industry performance across the globe. In addition, to stimulate domestic economic activities and enhance growth, the construction industry plays a role in the governments' effort [2], [3].

As growth globalized economy, the managing entire supply chain has become dynamic to the successful completion of a construction project. The strategy is very important to manage economic demand through construction to be more effective when projects are implemented innovatively and speedily [4]–[7]. However, the usage of IBS is still lower compared to others countries. In addition, a good SCM should be realized by the IBS players such as manufacturers, suppliers, developers, consultants, and contractors in order to enhance the performance of IBS construction approach [8]–[16]. Furthermore, understand the SCM is the major factor to improve the management and integration of supply chain [17], [18]. There is a problem related to delay in the construction industry which is a shortage of manpower, materials and variation order. Besides that,

Content from this work may be used under the terms of the Creative Commons Attribution 3.0 licence. Any further distribution of this work must maintain attribution to the author(s) and the title of the work, journal citation and DOI.
Published under licence by IOP Publishing Ltd

if the construction site is locating too far from the factory or suppliers, the contractors will have to bear some extra expenses for the transportation cost in the delivering product to the site.

According to [8], [9], [19], [20], the development of IBS in Malaysia is not well accepted by the parties for failure to keep the construction cost estimates in projects IBS. However, IBS can be seen and believed to reduce the number of challenges to improve this system, especially in structural elements the at Klang Valley [5], [16], [21].

The construction management of IBS project is important to ensure the successful implementation of IBS project. Therefore, this paper is mainly focusing on identifying the challenges and determine the issues in the SCM implementation of IBS in Klang Valley construction industry. This research is carried out by using a quantitative method which is questionnaires surveys in order to collect the data. The questionnaires are distributed to the IBS manufacturers who working in Klang Valley.

2. Methodology

To achieve the mentioned objectives, the methodology used in the preparation for this research was broadly divided into two (2) parts which are a literature review and questionnaire survey. A quantitative approach was employed using structured questionnaire survey to the manufacturers to gather the information and investigate the about the challenges and issues of the SCM on IBS in Klang Valley construction industry.

2.1. Literature Review

Seuring and Müller [22] defined a literature review is a systematic process that has a purpose which is to collect and analyze all relevant writings in the field by means of a structured search for literature. As a result, the conceptual content of the field can be identified and contributing to the theory development [23]. The literature review is an important element in the research because it describes how the proposed research is related to the prior research in statistics [24], [25]. It also shows the originality and relevance of the research problem. The literature review also justifies the proposed methodology and demonstrates the steps in preparing the completed research. The literature review is consisting of local and international referred journals, thesis, magazine, and websites [26]. Furthermore, the literature review provides information regarding the background of this research.

2.2. Questionnaire design

A pilot survey has been done and questionnaire is design based on the comments on the pilot questions. The number of questionnaires distributed to respondents was about 70 sets. The questionnaires were distributed by using google form and visits to offices and sites. The collection data of the main data will be done through it with the relevant parties involves on IBS in Klang Valley construction industry. All of the respondents have different perceptions about the challenges and issues that occur in SCM on the IBS in the construction industry.

This research has also involved the collection of data through semi-structured incorporated with interviews. All interviews are recorded and transcribed. The respondents of the interviews are selected on the basis of their experience on IBS. The questionnaire was structured into three sections; section A (demographic information of respondents' personal particulars in the construction industry, section B (the challenges of the implementation of the SCM in IBS construction industry and section C (The issues in the SCM in the construction industry). Questions in section B and C are designed based on Likert's scale of five ordinal measures scale of agreement for each statement which is from scale 1 to 5. In addition, the data collected need was analyzed using software such as SPSS Version 21 and interpreted in order to achieve the research objectives.

This paper is limited to the literature review and opinions from respondents that involved in the construction industry in Malaysian construction industry. There are no case studies or paper of Supply Chain Integration on IBS implementation in Klang Valley construction industry in the manufacturers' perspective which will provide a better result for the improvement of the SCM in this state if the challenges are identified.

3. Result and Analysis

This research focuses on the manufacturers perspective and a pilot research was conducted to ensure that the questionnaires are reliable and suitable for the research. The questionnaire survey was distributed randomly about 70 sets respondents from the companies in the Klang Valley, but only 35 of them had given the feedbacks with respond rate of 50%.

From the survey, it shows that majority of all them are engineers which have 45.7% followed by supervisors which have 31.4%, technician and others which are an operator, safety officer, and admin have 8.6% each, and the managers just 5.7%. The most respondent is from Engineer, which have 45.71%, while the rest are from managers, supervisors, technicians and others. The view from different position also helps in this paper because of the different job in the management of the construction industry on IBS. Besides that, the education level of the respondent's majority is a degree holder. The highest percentage of experience for the construction industry is 3 to 6 years, while the experience for the IBS is less than 3 years. This is because majority percentage of the respondents' regarding the age of the survey are from 31 to 40 years. In addition, the implementation of IBS is also contributory less to IBS experiences for major deficiencies.

The majority of all respondents know about the SCM with 71.43% and the rest 28.57%. Therefore, based on this data, the challenges and issues of supply chain integration on IBS implementation in Klang Valley can determine easily and reliable. Since majority all of the respondent know about the SCM, the respondents agreed that SCM is implemented at IBS. However, the challenges and issues should be determined to make sure the flow of the SCM can run smoothly.

3.1. The challenges of SCM on IBS implementation

Figure 1 is the analysis of the mean index for challenges of SCM in implement on IBS in the construction industry which are provided in the form of bar chart. Based on Figure 1, the result analysis shows the IBS requirement for a high level of coordination and integration of supply chains are from design, manufacturing, and construction stages are the highest challenge with the highest mean index which is 4.46, which is obviously majority all of the respondents agree. It means that the level of coordination and integration of supply chains of IBS requirement is high from each stage. This is due to the difficulties in working together and coordinating with their team which may have some conflicts in their scope of works.

Besides that, lack of communications between the manufacturer and IBS players will give the impacts on the quality of information and lack of skill and knowledge of the process and components of IBS will prevent interaction between IBS supply chain. Both have the same mean index which is 4.26 thus making these challenges second higher challenge of SCM in implement on IBS. Therefore, it shows that the respondents very agree with the challenges, this is because, the communication is a very important part of manufacturers and IBS players to avoid inaccurate design information, inaccurate data, and late updating of the required information. All of these problems could lead to the quality of the IBS delivery. The manufacturers themselves did not understand and familiar with the components and process of IBS. Therefore, this will hinder the interaction between IBS supply chain.

The mindset is more to the traditional role thinking can cause some challenges to supply chain on IBS adoption. The mean index is 4.14 which the respondents are agreed for this challenge. It shows that the rethinking of the old processes to move forward in the delivering of IBS from traditional management to SCM.

The customers demanding is a high priority in the supply chain implementation. It can be seen that mean index from the respondents' feedback is 3.91, which is represents the majority of all respondents agreed with this challenge. Thus, these result shows that their expectations are transformed towards a greater level of services which is increased profitability or reduced cost of customer order management performance.

Last challenge is the commitment from the seniors' managers to adapt this supply chain of IBS is still lacking. The mean index for this challenge is 3.89 which is all of the respondents agree. The seniors' managers were refused to switch from the conventional process. In addition, they refused to learn new things of management to their company unless it gives a benefit to the company regarding the cost.

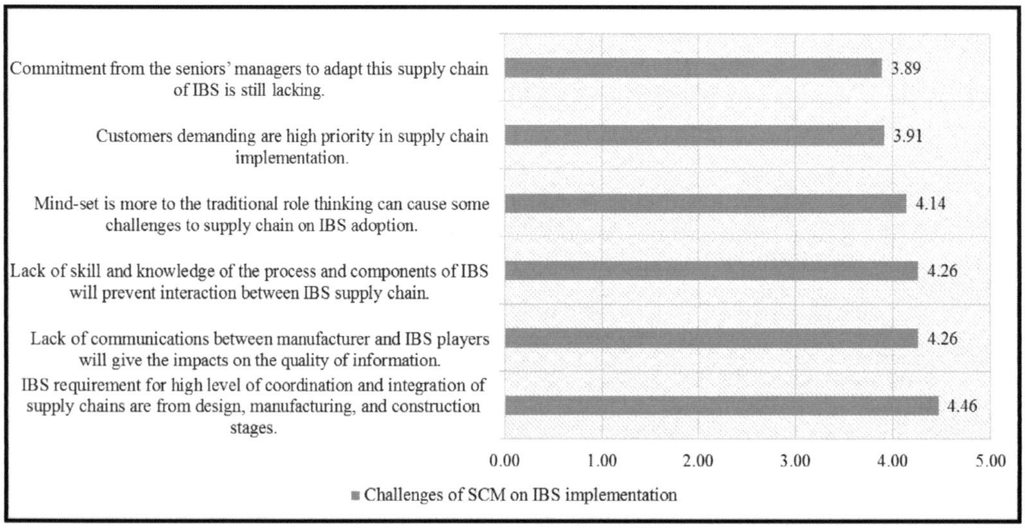

Figure 1. Mean Index Graph for challenges of SCM in implement on IBS

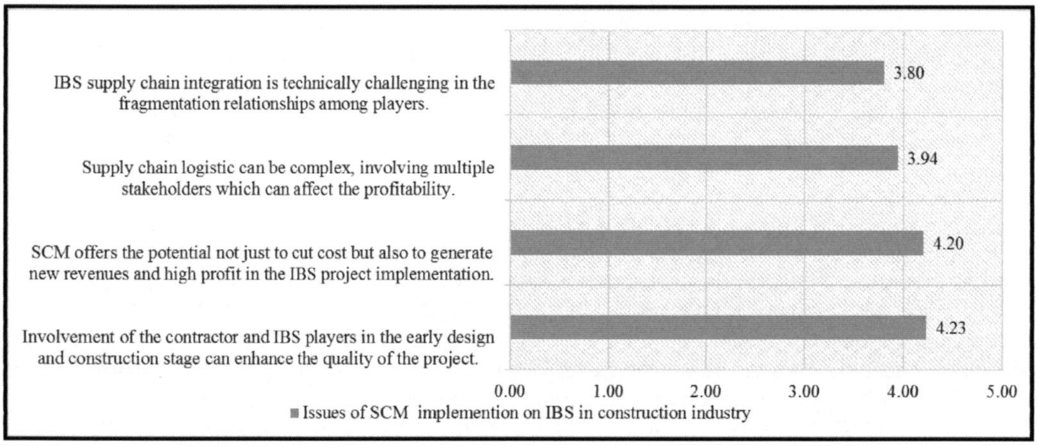

Figure 2. Mean Index Graph for Issues of SCM on IBS implementation

3.2. Issues of SCM on IBS implementation

From Figure 2, the result analysis shows that the highest issue is the involvement of the contractor and IBS players in the early design and construction stage can enhance the quality of the project with a mean index of 4.23, which all the respondents agreed. During the design stage, it is important if the manufacturers, contractors, and others IBS players involved in the design to prevent any crashes and errors. If there are any problems in the design, it will affect the SCM.

The second higher issues are SCM offers the potential not just to cut cost but also to generate new revenues and high profit in the IBS project implementation with a mean index of 4.20. The SCM could lead a high benefit which leads the cost reduction in the bank interest. The third higher issues are the supply chain logistic can be complex, involving multiple stakeholders which can affect the profitability with an average index of 3.94. The involving of multiple stakeholders could lead the fragmentation in design such as design clashes and errors, which it will affect the profitability and time consuming to redo the design. Finally, the IBS supply chain integration is technically challenging in the fragmentation relationships among players and the mean is 3.80 which is the respondents are agreed for these issues. This is due to the misunderstandings and misconceptions among the IBS players and manufacturers in working together could lead the IBS supply chain integration.

4. Conclusions

From this paper, the challenges on IBS implementation of SCM in Klang Valley are identified. Based on the respondents' feedback, the major challenges on IBS implementation of SCM that happened were IBS requirement for a high level of coordination and integration of supply chains are from design, manufacturing, and construction stages. The difficulties to work together of each team should not happen among manufacturers and IBS players. Next challenges are a lack of communications between the manufacturer and IBS players will give the impacts on the quality of information, and also a lack of skill and knowledge of the process and components of IBS will prevent interaction between IBS supply chain. Therefore, the communication skills and knowledge of IBS must have some improvement to ensure the delivery on IBS run smoothly to the site on time. These challenges are based on the respondents' experiences in the IBS construction industry.

Based on the respondents' feedback on the issues, the mean for all of the issues in Figure 8 is 4, where all respondents are agreed with the issues in the supply chain implementation of IBS. Therefore, the objective to determine the issues in the supply chain implementation of IBS in Klang Valley construction industry is achieved.

Some recommendation can be considered in SCM that the seniors' managers must have a commitment to adapt this SCM of IBS instead using of traditional management; the client should not change the designs during the construction stage. This is because, if the client does that, there will have some problems during the manufacture the IBS product, therefore it will cause the project delayed. There also must have a speaker who is expert about the SCM on IBS to introduce or give a talk about it, so that the IBS players such as manufacturers, clients, contractors, and others players can gain some knowledge about this system and have a confidences level to adapt the system.

Acknowledgement

The authors would like to express an acknowledgment to the Faculty of Civil Engineering, UiTM and to all the volunteers in completing the survey for this study. This research was funded by the Research Acculturation Grant Scheme (RAGS), Ministry of Education and Universiti Teknologi MARA, Malaysia.

References

[1] M. Qays, K. N. Mustapha, H. Al-Mattarneh, and B. S. Mohamed, "The constraints of industrialized building system from stakeholders' point of view," in IEEE International Conference on Semiconductor Electronics 2010, 2010.

[2] CIDB, "Industrialised building systems (IBS) - roadmap 2003-2010," 2003.

[3] CIDB, Construction Industry Master Plan Malaysia 2006-2015. 2007.

[4] V. Ahuja, "IT Enhanced Communication Protocols for Building Project Management by Small and Medium Enterprises in The Indian Construction Industry," Philosophy, pp. 1–401, 2007.

[5] A. S. A. Shukor, M. F. Mohammad, and R. Mahbub, "Supply chain integration challenges in project procurement in Malaysia: The perspective of IBS manufacturers," in 27 th Annual ARCOM Conference, 2011, pp. 495–504.

[6] M. F. Mohammad, A. S. A. Shukor, R. Mahbub, and F. M. Halil, "Challenges in the Integration of Supply Chains in IBS Project Environment in Malaysia," Procedia - Soc. Behav. Sci., vol. 153, pp. 44–54, 2014.

[7] A. S. A. Shukor, M. F. Mohammad, R. Mahbub, and F. Halil, "Towards Improving Integration of Supply Chain in IBS Construction Project Environment," Procedia - Soc. Behav. Sci., vol. 222, pp. 36–45, 2016.

[8] K. A. Kamar, M. Alshawi, and Z. A. Hamid, "Industrialised Building System: The critical success factors," in I9th International Postgraduate Research Conference, 2009, pp. 29–30.

[9] Z. A. Hamid, K. A. M. Kamar, and A. Mustafa, "The Critical Success Factors (CSFs) to IBS," Proc. IBS Int. Semin. Malaysian IBS Int. Exhib. 2009 (MIIE 2009), Kuala Lumpur, 21st – 23rd January 2009, vol. 2009, no. October 2003, pp. 21–23, 2009.

[10] K. A. M. Kamar, M. Alshawi, and Z. Hamid, "Barriers to Industrialized Building System (Ibs): the Case of Malaysia," Built Hum. Environ. 9th Int. Postgrad. Res. Confrence, no. 2009, pp. 1–16, 2009.

[11] K. Kamar and Z. Hamid, "Supply chain strategy for contractor in adopting industrialized building system (IBS)," Aust. J. Basic Appl. …, vol. 5, no. 12, pp. 2552–2557, 2011.

[12] M. K. Kamarul Anuar, A. H. Zuhairi, G. Mohd Khairolden, C. Egbu, and M. Arif, "Collaboration initiative on green construction and sustainability through Industrialized Buildings Systems (IBS) in the Malaysian construction industry," Int. J. Sustain. Constr. Eng. Technol., vol. 1, no. 1, pp. 119–127, 2011.

[13] K. A. M. Kamar, Z. A. Hamid, and N. Dzulkalnine, "Industrialised Building System (IBS) construction: Measuring the perception of contractors in Malaysia," in BEIAC 2012 - 2012 IEEE Business, Engineering and Industrial Applications Colloquium, 2012, pp. 328–333.

[14] Y.-S. Shin et al., "ScienceDirect Development of Internship & Capstone Design Integrated Program for University-Industry Collaboration," Procedia - Soc. Behav. Sci., vol. 102, pp. 386–391, 2013.

[15] N. Ernida et al., "The Strength Behavior of Self-Compacting Concrete Incorporating Bottom Ash as Partial Replacement to Fine Aggregate," Appl. Mech. Mater., vol. 774, no. February 2015, pp. 916–922, 2015.

[16] N. M. Nasir, M. N. M. Nawi, M. K. I. A. Rahim, A. Y. Bahaudin, and A. Tapa, "A review of delay factors in Malaysian industrialized Building System (IBS) construction project," ARPN J. Eng. Appl. Sci., vol. 11, no. 16, pp. 9868–9873, 2016.

[17] E. Scornavacca, E-Commerce and V-Business. 2007.

[18] S. B. Knouse and G. Fontenot, "Benefits of the business college internship: A research review," J. Employ. Couns., vol. 45, no. 2, pp. 61–66, 2008.

[19] F. Ismail, N. M. Yusuwan, and H. E. A. Baharuddin, "Management Factors for Successful IBS Projects Implementation," Procedia - Soc. Behav. Sci., vol. 68, pp. 99–107, 2012.

[20] C. Ezeah and C. L. Roberts, "Analysis of barriers and success factors affecting the adoption of sustainable management of municipal solid waste in Nigeria," J. Environ. Manage., vol. 103, 2012.

[21] K. A. M. Kamar, M. N. A. Azman, and M. N. M. Nawi, "IBS survey 2010: Drivers, barriers and critical success factors in adopting industrialised building system (IBS) construction by G7 contractors in Malaysia," J. Eng. Sci. Technol., vol. 9, no. 4, pp. 490–501, 2014.

[22] S. Seuring and M. Mueller, "From a literature review to a conceptual framework for sustainable supply chain management," J. Clean. Prod., vol. 16, no. 15, pp. 1699–1710, 2008.

[23] K. Burgess, P. J. Singh, and R. Koroglu, "Supply chain management: a structured literature review and implications for future research," Int. J. Oper. Prod. Manag., vol. 26, no. 7, pp. 703–729, 2006.

[24] S. J. Locke, L. F.; Spidurso, W. W.; Silverman, "The Function of the Proposal," In Proposals That Work. pp. 3–24, 2007.

[25] K. S. Taber, "Psychology for the classroom: constructivism and social learning," Teach. Dev., vol. 15, no. 4, pp. 555–557, 2011.

[26] J. P. Ollivier, J. C. Maso, and B. Bourdette, "Interfacial transition zone in concrete," Advanced *Cement Based Materials*, vol. 2, no. 1. pp. 30–38, 1995.

IOP Conf. Series: Materials Science and Engineering **291** (2017) 012016 doi:10.1088/1757-899X/291/1/012016

Time Overrun in Construction Project

Othman,I[1], Nasir Shafiq[1], Nuruddin,M.F[1]

[1]Civil and Environmental Engineering Department, Universiti Teknologi PETRONAS, 32610 Bandar Seri Iskandar, Perak, Malaysia.

E-Mail: idris_othman@utp.edu.my

Abstract. Timely completion is the key criteria to achieve success in any project despite the industry. Unfortunately construction industry in Malaysia has been labelled as industry facing poor performance leading to failure in achieving effective time management. As the consequence most of the project face huge amount of time overrun. This study assesses the causes of construction projects time overrun in Malaysia using structured questionnaire survey. Each respondent is asked to assign a one-to-five rating for each of the 18 time factors identified from literature review. Out of the 50 questionnaires sent out, 33 were received back representing 68% of the response rate. Data received from the questionnaires were analysed and processed using the descriptive statistics procedures. Findings from the study revealed that design and documentation issues, project management and contract administration, ineffective project planning and scheduling, contractor's site management, financial resource management were the major factors that cause the time overrun. This study is hoped to help the practitioners to implement the mitigation measure at planning stage in order to achieve successful construction projects.

Keywords: construction, time overrun, project, management, mitigation

1. Introduction

Construction industry is one of the highly dynamic sector and plays very important role in the development of any country. For that, time or schedule is the fundamental criteria for the success of any projects in the world regardless the industry [1]. Malaysia is a fast developing country in the Asian region and has gone through such rapid economic growth since the seventies [2]. The industry itself has been consistently contributed approximately 3% to 5% of the national Gross Domestic Product (GDP). Basically there are two main sector for construction projects in Malaysia namely public and private sector [3]. Most of the public sector projects are managed by Public Works Department (PWD). In Malaysia, the Construction Industry Development Board (CIDB) is a body that function as developing, improving and expanding the Malaysian construction industry and is involved with the public and private sectors project development [4].

1.1. Background of Study

The objective of a project itself is to produce a quality project, completed within time frame provided, within sufficient budget and in a safe working environment. Normally, in the construction industry, the aim of project control is to ensure all the projects could be completed on time [5]. A successful project

Content from this work may be used under the terms of the Creative Commons Attribution 3.0 licence. Any further distribution of this work must maintain attribution to the author(s) and the title of the work, journal citation and DOI.
Published under licence by IOP Publishing Ltd

is the only project which has been accomplished its technical performance, maintained the schedule, and remained within budgetary costs. The project management tools and also the techniques play an important role in the effective management of any projects [6]. The construction time is so much fundamental consideration in the project management and regarded as most important factors for indicate the successes of any projects. Poor performance in terms of time will lead to the significant overrun which is global phenomenon nowadays [3].

1.2 Problem Statement

In any construction project, one of the main goal of practitioners is to achieve timely completion of projects within its allocated budget and required quality as each day of time overrun in the making process of any project has direct impact on the cost of the project. In Malaysia, traditional lump sum payment system, design and built system and Construction Project Management/Contract Management are commonly adopted in the procurement strategies [4]. However, based on any literature review, despite adopting various management practices, construction projects in many countries are still facing problem such as time overrun which really needs serious attention. Construction industry such as in Malaysia is facing chronic problems including poor performance of time and cost, construction waste, poor productivity and over dependent so much on foreign workers. All of these problems, poor in time and cost performance are considered as a critical issue [6]. Aibinu [2] reported that 7 out of 10 projects surveyed in Nigeria suffered the issue of delays. 70% of the large construction projects studied in the Saudi Arabia experienced average time overruns between among 10% to 30%. 50% of construction projects in United Arab Emirates (UAE) encountered delays. During the last few decades, numerous project control methods have been developed such as Gantt Bar Chart, Program Evaluation and Review Technic (PERT) and Critical Path Method. Variation of software have also been introduced to assist project control methods such as Microsoft Project, Asta Power Project, Primavera. Despite the use of these methods and software in practice, and yet still many construction projects suffer time overruns [7].

1.3 Objectives and Scope of Study

This research is mainly focused on three objectives; identify the critical factors which lead to time overrun in construction project, to recognized and identify how these factors could occur and to propose mitigation measures on how to reduce number of time issues. This research and review focuses on the factors of time overrun at construction project in Malaysia. Correspondently, this research will also determine the most factors causes' time cost overrun and rank them, based on the most chosen by the respondents.

2. Literature Review

In general, Malaysian construction industry itself is categorized into 2 areas as first category is the general construction, consist of residential construction, non-residential construction and civil engineering construction. Second category named as special trade works which comprises activities of metal works, plumbing, sewerage and sanitary works, refrigeration and air conditioning works, painting works, carpentry, flooring works, tiling and glass works. A construction project is a project that relate to any of these areas in the construction industry [4]. According to Endut [3], only 30% of construction projects were completed within specific scheduled dates and the average time overrun was between 10% and 30%. Looking into detail of Malaysian construction projects, it is found that there is a significant growth in the construction industry throughout the years. According to Market Watch Report 2012, the construction sector strengthened further by 6.3% during the first half of 2010 which is from January 2009 till June 2009 with 2.9%. The expansion was largely due to increase of civil engineering and non-residential activities. The rapid implementation of construction projects under the Ninth Malaysian Plan (9MP) and stimulus package provided by the government contributed largely in this expansion. The construction sector expanded further up till 3.5% in the year 2011. Time overrun could be defined as late of completion of works as per compared with planned schedule or contract schedule. It occurs when the progress of job done behind its fixed schedule [1]. Other than that, construction project time overrun can also be defined as an extension of time beyond the contractual agreement during tender process. It may have adverse effects on both the owner and contractor (either in the form of lost revenues or extra expenses) and it is often raises the contentious issue of delay responsibility, which later may result in the conflicts that frequently reach the courts. Since the problem of project delay some sort of contextual, studies have been conducted to investigate this problem in the construction industry in several countries. A lot of researches have outlined the issue of poor time performance of the construction projects globally [6]. There have been studies of 8000 projects where the researchers found that only 16% of the projects could satisfy the three famous performance criteria namely completing projects within time frame, within budgeted cost and also quality standard. Malaysian construction industry also facing these problems significantly. In a survey of delay practices in most Malaysian construction industry, the result stated that 87% of the respondents reported that they have encountered delays in projects with overrun in time of 10-40% of contract. Furthermore, study of MARA large projects that concluded the construction projects encountered an overrun with average of 23.47% of contract duration [3].

3. Methodology

This research is carried out in two stages whereby firstly, literature survey and the interview with the project superintendent working for the Public Works Department in Perak Tengah are conducted to determine the factors influencing the time and cost overrun in construction and oil and gas projects. Eighteen factors are identified and categorized into two different groups in this research. These groups or sections shall cover partial of the criteria that need to be considered in the time factors concern. Secondly, a questionnaire survey methodology is employed to determine and rank these factors based on their levels of importance on the time overrun. The survey questionnaires are sent via gizmo survey – online survey form – to seniors and known associates. The questionnaires also are sent via hand delivery to 50 respondents either engineer, superintendent officer, quantity surveyor and interns. The response rate for the questionnaire survey is about 68% which is very much higher than the normal rate of 20% and 30% for most postal questionnaire surveys of the construction industry. The analysis of survey is done by Relative Importance Index (RII) method. The scores are then transformed to importance indices based on the following formula:

$$\text{Relative importance} = \frac{\sum w}{AN}$$

Where w is the weighting given to each factor by the respondents, ranging from 1 to 5, A is the highest weight (i.e. 5 in the study) and N is the total number of samples. At the same time, the analysis also was based on the qualitative measurement or ranking system. Rating for the questionnaire is 1 – Totally Disagree, 2 – Disagree, 3 – Moderately, 4 – Agree, 5 – Totally Agree. The Average Index Formula as follow [8]:

$$\text{Average Index (AI)} = \sum(\beta \times n)/N$$

Where, ß is weighing given to each factor by respondents, n is the frequency of the respondents, N is the total number of respondents.

4. Result and Discussion

The questionnaires had been distributed to the professional in construction projects and government agency. A total of 50 questionnaires for the total of industry are distributed through self-delivery of hardcopy questionnaires to the companies involved. The questionnaire consists of three main sections, General Information, Likelihood of Factors of Time Overrun to Occur and Current Practice of Time Cost Management in the Construction Project. From the survey, eighteen (18) respondents working in projects have Bachelor's Degree. Three (3) respondents have Master's Degree and four (4) respondent has diploma as their lowest education. Indeed, from all 33 respondents, none of the have PhD for their highest education. Meanwhile the rest have other lower qualifications such as Sijil Pelajaran Malaysia (SPM), Technical Certificate or Diploma as their highest education. Based on the data collected above, the author has extracted the most crucial which is top 10 factors that would be the factors toward the time overrun and author has ranked the questions based on its relative importance index as shown in Table 1. There is one factor that possess RII of 0.915 (closest to 1.00) while the lowest among top 10 out of 18 has the RII value of 0.800 which was considered as the least important information for this research. The author has arranged the Likelihood of Factors of Time Overruns to Occur based on the ranking order as shown above. This indicates that complexities and non-performance of subcontractors is not included in the factors of time overrun.

Table 1. Summary of Top 10 Time Overrun Factors

No.	Factor of Time Overrun	RII	Rank
1	Design and Documentation Issues	0.915	1
2	Project Management and Contract Administration	0.892	2
3	Contractors Site Management	0.862	3
4	Ineffective Project Planning and Scheduling	0.854	4
5	Financial Resource Management	0.831	5
6	Material and Machineries Resource	0.815	6
7	Issues regarding permissions/approvals	0.815	7
8	Design Changes	0.808	8
9	Inaccurate Evaluation of Project / Time Duration	0.800	9
10	Complexities and Non-Performance of Subcontractors	0.800	10

All the result from Table 2 shows a high range number which is between 3.40 and 5.00. This can be indicated that the respondents were neutral/agreed/strongly agreed with the statement given by the author for the most 2 section, B and C from the questionnaire. Therefore, based on the Average Index (AI) data, the questions from the survey can be considered as the valid of set of questions for time and cost overrun factors.

Table 2: Average Index for Factors of Time Overrun

No.	Factor of Time Overrun	AI	Rank
1	Design and Documentation Issues	0.915	1
2	Project Management and Contract Administration	0.892	2
3	Ineffective Project Planning and Scheduling	0.862	3
4	Contractors Site Management	0.854	4
5	Financial Resource Management	0.831	5
6	Design Changes	0.815	6
7	Inaccurate Evaluation of Project / Time Duration	0.815	7
8	Issues regarding permissions/approvals	0.808	8
9	Complexities and Non-Performance of Subcontractors	0.800	9
10	Material and Machineries Resource	0.800	10

Cronbach's Alpha value is about 0.899 and 0.884 as shown in Table 3 which author can consider as good internal consistency. High value indicate that the information gathered are reliable and consistent so that it could not be questions by others. Some of the questions if were deleted, will provide higher consistency of Cronbach's Alpha.

Table 3: Reliability Statistics for Time Overrun Factors

Cronbach's Alpha	Cronbach's Alpha Based on Standardized Items	N of Items
0.899	0.894	18

4.1 Analysis of Top 5 Time Overrun Factors

Based on the survey conducted and through the analysis process, the most likely factor of time overruns to occur is Design and Documentation Issues with relative importance index (RII) of 0.915. Highest RII indicate that as per mentioned above nearest to 1 show that this factor affecting the time overrun for most of the projects. From the structured interview done with Engineer of JKR Perak Tengah, Mr Faizal, he mentioned that from the point of view as the owner, in construction project implementation process, there are few stages such as feasibility study, planning, technical design stage, and so on whereby those stages done by different team and different department which really make this factor the most likely to occur.

The second factor most likely to occur is Project Management and Contract Administration with RII of 0.892. This factor affects the overall progress of the project in terms of optimization of time. Based on the interview, this factor reflects on how the project manager handle the case. For example, if the person experts with the project, it should be non-issue since he knows how to deal with. It is also depending on how the project manager keep on discussion with team members for any new projects.

Contractors Site Management ranks at 3 among the all other top Time Overrun Factors. The number of RII after analysed it is about 0.862 and this also indicate that it is really related with the time overruns for construction and might be Oil and Gas Industry. In any construction project, 3M such as manpower, money and materials involves directly towards the betterment of the income. Synchronized availability of those 3M's can only lead to the progress at site. Site management also affected based on 3M as the contractors already agreed with the number of workers, amount of money, material ordered, but at the end of the day, all of the would have effect of it.

3M as mentioned before also related with this factor. The planning and scheduling done based on the number of manpower, money and material that contractor have. So it is very crucial to optimize each task with each manpower so that it great outcome can be achieved. Cash flow in the construction project should be done carefully at any time by the project manager and arrange it necessary at requisite time helps the project to move as per schedule. Nowadays, a lot of software have been created to ease the task, but at the end of the day, it is still depending of the project manager competency to cope with the issues.

This factor related with time overrun where the contractor should have particular within a period, such as concreting process, but it not delivered due to the payment or bad cash flow. Hence, the time of concreting shall be delayed until the contractor done the payment to the batching plant. On behalf of the person who shall delivered the payment, a few stages need to go through and endorsed by a few persons, and in that sense of manner, we would see the delay of payment occur. Policy of government mentioned that progress payment shall be done within 14 weeks starting from the certifying the amount of progress of the projects.

5. Conclusion

The main objectives of this research are to identify the issues and factors that faced by the contractors, consultant and clients due to the time overrun, investigate the causes of such issues and to propose some suggestion for future performance in term of time effectively. Results from this research are based on the respond from the respondent which already being distributed during this research. Besides, at the end of this research the author would like to emphasis on proper time management control in construction project as it would assure the effectiveness of a project. Based on research done, improvements in many aspects need to be considered and monitored frequently in order to ensure the relevancy of time performance in construction project. Mitigation measures that can be applied are external factors such as under staff should not be an issue unless the company already applied to have more but not yet success, they need to have a good plan to utilize and optimize all the staff. Project manager should be more tolerance and competency. Competency here is about more to communication and thinking skill. Indeed, we deal with person which have various background hence it is important for him or her to be flexible and can cope with the tasks given.

References

[1] Memon, A. H., Rahman, I. A., Abdullah, M. R., & Aziz, A. A. A. (2011). Time Overrun in Construction Projects from the Perspective of Project Management Consultant (PMC). Journal of Surveying, Construction & Property. Volume 2, Issue 1, University of Malaya, Malaysia. 54-66.

[2] Aibinu, A. A., & Jagboro, G. O. (2002). The Effects of Construction Delays on Project Delivery in Nigerian Construction Industry. International Journal of Project Management, 20(8), 593-599.

[3] Endut, I. R., Akintoye, A., & Kelly, J. (2005). Cost and Time Overruns of Projects in Malaysia. Proceedings of the 2nd Scottish Conference for Postgraduate Researchers of the Built and Natural Environment 16-17 November 2005,Glasgow Caledonian University, 243-252. (2007).

[4] CIDB Construction Industry Master Plan (CIMP) 2006-2015.

[5] Ahmad, S. A., Hassan, F., Hassan, S., Mat, M. C., Nasir, N. M., & Samad, Z. A. (2009). A Study on the Practise of Delay Analysis Techniques in the Malaysian Construction Industry. Paper presented at the 13th Pacific Association of Quantity Surveyors Congress (PAQS).

[6] Memon, A. H., Rahman, I. A., & Azis, A. A. A. (2012). Time and Cost Performance in Construction Projects in Southern and Central Regions of Peninsular Malaysia. International Journal of Advances in Applied Science.

[7] Olawale, Y. A., & Sun, M. (2010). Cost and Time Control of Construction Projects: Inhibiting Factors and Mitigating Measures in Practice. Construction Management & Economics, 28(5), 509-526.

[8] Majid, M. Z. & McCaffer, R. (1997). Discussion of Work Performance of Maintenance Contractors in Saudi Arabia. Journal of Management in Engineering ASCE, Vol. 13, No. 5, Pg. 91.

Quality planning in Construction Project

Othman,I[1], Nasir Shafiq[1], Nuruddin,M.F[1]

[1]Civil and Environmental Engineering Department, Universiti Teknologi PETRONAS, 32610 Bandar Seri Iskandar, Perak, Malaysia.

E-Mail: idris_othman@utp.edu.my

Abstract. The purpose of this paper is to investigate deeper on the factors that contribute to the effectiveness of quality planning, identifying the common problems encountered in quality planning, practices and ways for improvements in quality planning for construction projects. This paper involves data collected from construction company representatives across Malaysia that are obtained through semi-structured interviews as well as questionnaire distributions. Results shows that design of experiments (average index: 4.61), inspection (average index: 4.45) and quality audit as well as other methods (average index: 4.26) rank first, second and third most important factors respectively.

Keywords: quality planning, construction, projects, design, audit

1. Introduction

The concept of quality planning in construction projects is to guarantee efforts to accomplish the necessary level of quality for the outcome which are well planned and organized. It is vitally required for a construction company to have an effective quality management system as it helps organizations in improving customer satisfaction whilst providing the organization with a competitive advantage over fellow competitors within the industry [1]. It is about obtaining customers' satisfaction that would lead to long term competitiveness and business survival for the companies by maintaining the quality of construction activities at a mandatory standard.

1.1. Background of Study

The term quality is defined as fitness for use [2]. It is about achieving or going beyond customer expectations at a price that signifies a value to them. Generally, quality project planning is to produce the deliverables of the project successfully using the establishment of asset of directions in sufficient fact to inform the project team exactly what must be done, when it must be done and what resources to use in [3][4].

1.2 Problem Statement

As observed throughout the years, the achievement of adequate levels of quality in the construction industry has been a setback due to non-existent and poor quality of planning practices, a lot of the construction projects do not achieve their targets [5],[6] where by major outflows of time, money and resources, both human and material are wasted each year[7],[8]. As a result, the need for improvement in project success in the construction industry is the main motivation for this research. This study focuses

Content from this work may be used under the terms of the Creative Commons Attribution 3.0 licence. Any further distribution of this work must maintain attribution to the author(s) and the title of the work, journal citation and DOI.
Published under licence by IOP Publishing Ltd

IOP Conf. Series: Materials Science and Engineering **291** (2017) 012017 doi:10.1088/1757-899X/291/1/012017

particularly on the quality planning stage of a project, due to its high magnitude in determining project success.

1.3 Objectives and Scope of Study
The objectives of the research are to identify factors that contribute to the effectiveness of quality planning in construction project. Investigate problems that are encountered in quality planning for construction project and determine the improvements that can be implemented for quality planning in construction project.

2. Literature Review
In general, Quality is the characteristic element of an item that can be evaluated as a meeting standard, whereby if it meets or exceeds the standard, it can be said to be of good quality or high quality. However if the item does not meet the standard, it is considered poor quality [9]. Quality Planning is about identifying which standard is relevant to the project and determining how to satisfy them [10].

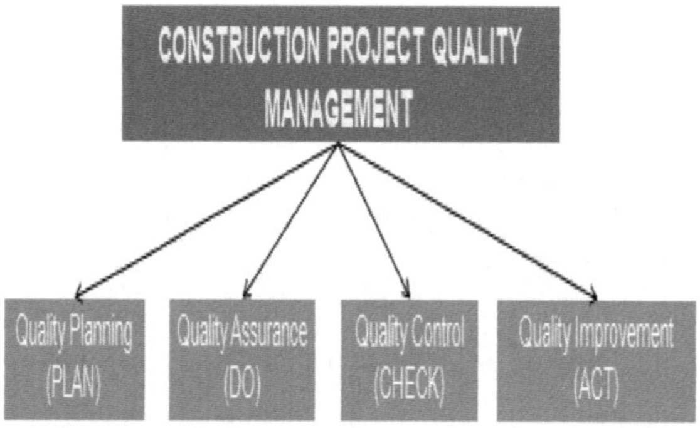

Figure 1. Approaches to conformance in Quality Management [11]

Based on the Figure 1 above, there are four basic approaches identified in Quality Management, which are the quality planning, quality control, quality assurance and quality improvement [11]. Quality Planning is basically one of the steps in Quality Management. In order to proceed with the do, check and act, it is essential to prepare quality plan ahead of time in order to produce quality outcome. Quality management provides the environment within related procedures, techniques and tools that can be deployed effectively leading to the project success [12].

3. Methodology
This research is carried out in two stages whereby firstly, factors from literature review were identified. Secondly, a questionnaire survey methodology is employed to determine and rank these factors based on their levels. The survey questionnaires are sent via online survey form to seniors and known associates. The questionnaires also are sent via hand delivery to 50 respondents either engineer, superintendent officer, quantity surveyor and interns. The response rate for the questionnaire survey is about 68% which is very much higher than the normal rate of 20% and 30% for most postal questionnaire surveys of the construction industry. The analysis of survey is done by Relative Importance Index (RII) method. The scores are then transformed to importance indices based on the following formula. The analysis also was based on the qualitative measurement or ranking system. Rating for the questionnaire is 1 – Totally Disagree, 2 – Disagree, 3 – Moderately, 4 – Agree, 5 – Totally Agree. The Average Index Formula as follow [13]:

$$\text{Average Index (AI)} = \Sigma\,(ß \times n)/N$$

Where, ß is weighing given to each factor by respondents
n is the frequency of the respondents
N is the total number of respondents

4. Result and Discussion

The questionnaires had been distributed to the professional in construction projects and government agency. A total of 50 questionnaires for the total of industry are distributed through self-delivery of hardcopy questionnaires and email to the companies involved. Figure 2 shows the respondent's designation which consist of project manager, executive director, quality system manager, management systems manager, project coordinator, assistant general manager and project engineer. Respondent's designation category is much important in this research as it adds validity to set opinions and ideas of respondents on problems in quality planning in construction project in Malaysia. Project managers have the highest proportion (33.3%) and lowest are Executive Director, Management Systems Manager and Assistant General Manager (6.67%).

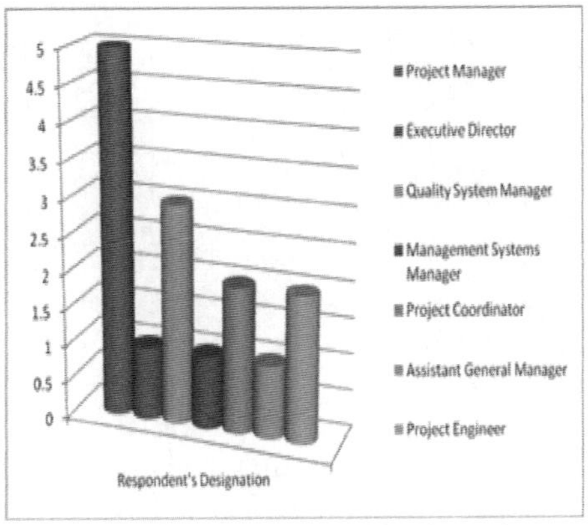

Figure 2. Respondent Designation

Figure 3 illustrates the area of experience in quality planning for respondent. Having been allowed to identify the perspective with one or more quality planning functions, the respondents showed a relatively uniform representation of the functions in quality plans, with design of experiment and inspection as highest areas of experience. Quality audit and other methods as the second most common areas of expertise among the respondents.

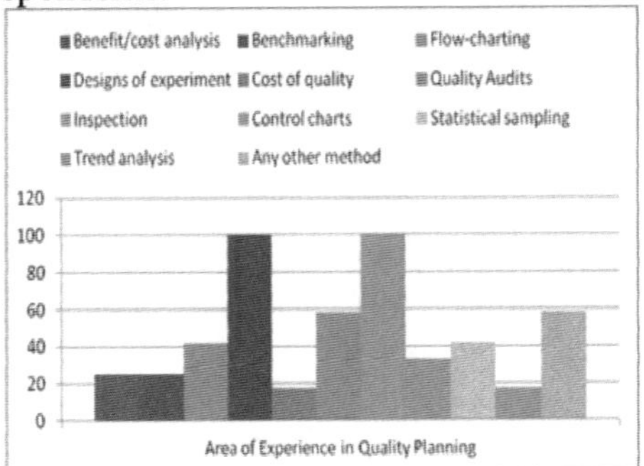

Figure 3. Area of experience in Quality Planning

Table 1 shows the ranking for each of the aspect included in the questionnaire. The aspect ranking is analysed by using Average Index Formula method.

IOP Conf. Series: Materials Science and Engineering **291** (2017) 012017 doi:10.1088/1757-899X/291/1/012017

Table 1. Important Aspect in Quality Planning

Rank	Important Aspects	Average Index
1	Benefit/cost analysis	4.01
2	Benchmarking	4.13
3	Flowcharting	4.1
4	Design of Experiments	4.61
5	Cost of Quality	4.06
6	Quality Audits	4.26
7	Inspection	4.45
8	Control Charts	4.24
9	Pareto Diagram	4.03
10	Statistical Sampling	4.24
11	Trend Analysis	4.09
12	Any Other Methods	4.3

The important factors or aspects in quality planning agreed by respondents are the design of experiments (average index: 4.61), inspection (average index: 4.45) and quality audit as well as other methods (average index: 4.26) rank first, second and third most important aspect respectively. It is expected that design of experiments (DOE) ranks first because it is an essential method in quality planning. DOE is the design of any information-gathering exercises where variation is present, whether under the full control of the experimenter or not. It should be simultaneous with the project objectives and goals to have an effective integration with the project. If DOE is not carried out properly, it can create problems during the execution of the project. Inspection which is the next most important aspect from the ranking is so significant in quality planning because inspection is one of the most important elements in achieving a successful construction project. Proper inspection requires sound judgment, diplomacy, procedural knowledge, effective inspection techniques, and complete and accurate documentation. Quality audit is also another important aspect in the quality planning because it affects the project flow and systems in different ways. Quality audit is basically the process of systematic examination of a quality system carried out by an internal or external quality auditor or an audit team. It is an important part of organization's quality management system and is a key element in the ISO quality system standard, ISO 9001.

However, the least important aspect from the ranking are pareto diagram (average index: 4.03), Cost of Quality (average index: 4.06) and Trend Analysis (average index: 4.09). Probably, pareto diagram is insignificant to them for some reason. One of the reason might be because it is a type of chart that contains both bars and a line graph which is not accurate enough to contribute in quality planning. Inaccuracy can result in decrease in chances of achieving a quality project execution. Cost of quality is among the bottom ranked aspect as it is the general perception, that higher quality requires higher costs, either by buying better materials or machines or by hiring more labour where by in current projects, the cost are basically limited and fully dependant on the country's economy causing the priority of extra expenditures to be less prioritized. Trend analysis is rank as number 10 because most of the companies prefer their own personal methods and project flow, rather than analyse trends of other projects.

5. Conclusion

The purpose of this study was to identify important aspects or factors that contribute to the effectiveness of high-quality planning construction projects. The study was conducted through survey questionnaires, and semi-structured interviews with project team members in Malaysia. The research results showed that Malaysian construction companies used in the study exposed to light different approach to quality. The results indicated several barriers to planning for quality, including a lack of quality enterprise culture and the difficulty of developing a plan for long-term quality. The study also showed that most companies are reluctant to change their current practice because clients are not driving a high quality approach based on the selection of the contractor. The observed results indicate that the economic crisis that disrupted over the development of a quality plan. Generation of the results of this study, the following conclusions can be made for the construction industry. The planning process identified the common quality on the basis of the study will be useful for other contractors, subcontractors especially the lower graduates who do not have a quality system in practice. Therefore, it is important to conduct training programs to improve the knowledge of Malaysian entrepreneurs in these areas and approach to quality planning feasible and attractive enough to participate in the larger market. The concept of quality planning should still be educated, and a process that is suitable for the construction industry while overcoming the obstacles identified should be developed to achieve better results with investing in quality.

References

[1] Meredith, J. R. and Mantel, S. J. (2006), Project Management – A Managerial Approach, 6th ed.,Wiley, New York, NY.
[2] Harris, F., & McCaffer, R.(2001). Modern Construction Management (5th ed.). Oxford: Blackwell Science Ltd.
[3] Turner, J.R. (1999), The Handbookof Project-based Management: Improving the Processes for Achieving Strategic Objectives, McGraw-Hill, London.
[4] Leung, M. Y., Ng, S. T. and Cheung, S. O. (2004), Measuring Construction Project Participant Satisfaction, Construction Management and Economics, Vol. 22 No. 3, p.319.
[5] Kerzner, H. (2006), Project Management: A Systems Approach to Planning, Scheduling and Controlling, 9th ed., Wiley, New York, NY.
[6] Robinson, H. S., Carrillo, P. M., Anumba,C. J. and Al-Ghassani, M. (2005), Review and Implementation of Performance Management Models in Construction Engineering Organisations, Construction Innovation, Vol. 5 No. 4, pp. 203-217.
[7] Tan,C. K., & Abdul-Rahman, H. (2005). Preliminary Research into Overcoming Implementation Problems in Construction Projects. Proceeding of the 4th Micra Conference. Faculty of the Built Environment, University Malaya, pp. 08.15-08.28.
[8] D.A.Bumb, D .M. Ghaitidak (2016), Study of Quality Assurance and Quality Management System in Multi storey RC Building, International Journal on Recent and Innovation Trends in Computing and Communication ISSN: 2321-8169 Volume: 4 Issue: 5, pp. 112 – 113.
[9] Sarshar, M., Haigh, R. and Amaratunga, D. (2004), Improving Project Processes: Best Practice Case Study, Construction Innovation, Vol. 4 No. 2, pp. 69-82.
[10] Biggar, J. L. (1990). Total Quality Management in Construction. Transactions of the American Association of Cost Engineers, Q.1.1-Q.1.4.
[11] Dissanayaka, S. M., Kumaraswamy, M. M., Karim, K., & Marosszeky, M. (2001). Evaluating Outcomes from ISO 9000 - Certified Quality Systems of Hong Kong Constructors. Total Quality Management, 12(1), pp. 29-40.
[12] Abdul-Rahman, H. (1996). Some Observations on the Management of Quality among Construction Professionals in the UK. Construction Management and Economics, 14, 485-495.
[13] Majid, M. Z. & McCaffer, R. (1997). Discussion of Work Performance of Maintenance Contractors in Saudi Arabia. Journal of Management in Engineering ASCE, Vol. 13, No. 5, Pg. 91.

Effective Safety Management in Construction Project

Othman,I[1], Nasir Shafiq[1], Nuruddin,M.F[1]

[1]Civil and Environmental Engineering Department, Universiti Teknologi PETRONAS, 32610 Bandar Seri Iskandar, Perak, Malaysia.

E-Mail: idris_othman@utp.edu.my

Abstract. Effective safety management is one of the serious problems in the construction industry worldwide, especially in large-scale construction projects. There have been significant reductions in the number and the rate of injury over the last 20 years. Nevertheless, construction remains as one of the high risk industry. The purpose of this study is to examine safety management in the Malaysian construction industry, as well as to highlight the importance of construction safety management. The industry has contributed significantly to the economic growth of the country. However, when construction safety management is not implemented systematically, accidents will happen and this can affect the economic growth of the country. This study put the safety management in construction project as one of the important elements to project performance and success. The study emphasize on awareness and the factors that lead to the safety cases in construction project.

Keywords: construction, safety, injury, awareness, management

1. Introduction

According to Ahmadon [1] construction sites are generally complex and sometimes unsafe to the workers and environment. It is one of the most dangerous workplaces because of high number and frequency of accidents. Therefore, construction safety is one of the serious problems in the construction industry worldwide, especially in mega construction projects. This is because of the involvement of many workers, modern methods of construction, many large and heavy plants, a great amount of materials and equipment used, complex construction operation, multi-interface, and various disciplinary aspects of its project workforce. All of this criteria that lead to the higher accident rate at the construction projects. The accidents include those that result from falling from height, collision, collapse and electric shock, amongst which falling from height and collisions are the most prevalent [2].

1.1. Background of Study

Construction safety as a result continues to represent a serious problem and pose a challenge for researchers and practitioners [3]. In Malaysia, both the society and economy have suffered human and financial losses as a result of the poor safety performance in the construction industry. Department of Occupational Safety and Health (DOSH) in Malaysia reports that occupational accidents by sector shows that the highest number of death was in the construction industry as of for the 2015 incidents [3].

Content from this work may be used under the terms of the Creative Commons Attribution 3.0 licence. Any further distribution of this work must maintain attribution to the author(s) and the title of the work, journal citation and DOI.
Published under licence by IOP Publishing Ltd

IOP Conf. Series: Materials Science and Engineering **291** (2017) 012018 doi:10.1088/1757-899X/291/1/012018

1.2 Problem Statement

Effective safety management on project sites is of utmost importance due to the nature of the construction industry. However, it is usually not as top priority in a market-driven society where the main focus is completing projects at the specific quality with minimum time and cost. Thus, safety issues are considered only after an accident occurs at a construction site. Therefore, proper effective safety management in construction is of utmost importance. Various factors that affecting construction safety performance were explored through a questionnaire survey responded by people that involve in construction projects.

1.3 Objectives

This project focuses on few objectives, which are:

i. To study the safety issues that happen in construction projects
ii. To identify variables that significantly affects construction safety

1.4 Scope of Study

The scopes of study in this project as below:

i. Safety policy at construction projects
ii. Factors that lead to safety cases
iii. Method that applied to reduce the frequency of accidents at construction sites

2. Literature Review

Effective safety management in construction projects is a core consideration for all types of organizations that are accountable for protecting and optimizing the efficiency of human resources. In regard to construction, ensuring workplace safety is not an easy task. Occupational accidents in the construction industry will give impact to economic and social issues in organizations, as well as countries. Abdul Rahim [4] stated that the development of construction industry has been plagued by the accidents or injuries that are frequently occurred. It is estimated that there are around 60,000 construction fatalities occurred worldwide each year, which equates to one accident happens every nine minutes. Among all industries in the world, construction has the highest rate of accidents, including deaths and disabling injuries [5].

Many ways have been made to minimize this problem, but the outputs are far from satisfactory as construction accidents continue to dominate. Despite various programs implemented by government authorities at a national level and the initiatives of private companies, the number of construction accidents remains alarmingly high. It is evident that these efforts are not sufficient to minimize the occurrence of unsafe acts and unsafe conditions at construction sites. According to Chang [6], accidents at work occur either due to lack of knowledge, supervision, lack of means to carry out a work safely, errors in judgment, carelessness in making decision, or total irresponsibility. In addition, the lack of a controlled working environment and the complexity and diversity of the sizes of organizations have given impacts on safety performance in the industry. Inadequate safety precautions, non-implementation of rules, deficient knowledge and unqualified officers cause unexpected accidents in the construction project. Furthermore, the safety practices adopted at construction sites are far below acceptable standards.

According to Edward [7], the safety legislation and policies have a great impact upon the safety level of a construction worksite. Legislation forms a framework in which health and safety is regulated and controlled. In this case, management have to follow the rules and regulations duly and punishments to be meted out to those who flout them. The identification of unsafe factors in the construction field mainly highlighted on the site layout, multi-interface, safety screen and scaffolding, plant operation, and construction operation [8]. The identification processes are based on type of work conducted at the sites, which integrates the safety risks. By walking through the site and inspecting all the workers work, the managers can identify the potential hazard factors by themselves.

3. Methodology

This research was conducted to establish effective safety management in construction project. It also will act as a benchmark to analyze the safety policy and the factors that lead to the safety accidents at construction sites. The research can be divided into phases as shown in Figure 1 which are from preliminary study phase, data collection phase based on the questionnaire, data analysis phase and conclusion. A questionnaire was designed with the objective of determining the variables that affect safety management in construction.

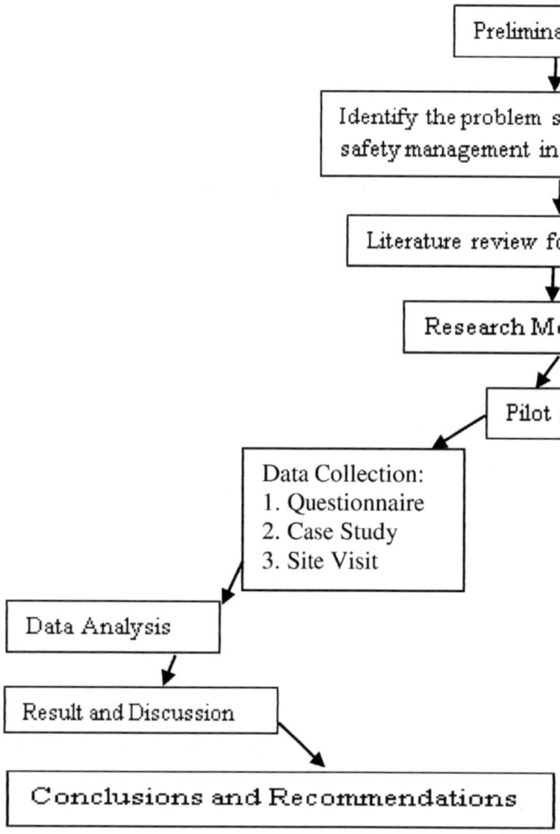

Figure 1. Project Activities Flow Diagram

4. Result and Discussion

The pilot survey had been done throughout all of the professional engineers either in the academic, construction or consultancy industry. After obtaining and analysing the results of the pilot survey, logistical, technical and other issues or problems have been addressed. The questionnaire format was revised and the type of survey had been altered into a more suitable one. After the revision of the survey being made, the large-scale of the survey was executed.

The questionnaires had been distributed to the project teams which base at Bangi area. It consists of the client, consultant, contractor and companies. Since this study is a qualitative study, 30 feedbacks are sufficient enough. The questionnaire consists of two sections, General Information and Factors Affecting Effective Safety Management in Construction Project. Based on the Average Index and Relative Importance Index (RII) as shown in Table 1, Figure 2 and 3, the respondents rank the first factor affecting effective/improper safety management in construction is "Safety training and awareness", with a relative importance index of 0.91 and average index of 4.57. It indicates that safety training and awareness play a very important role in safety management in construction projects. "Worker's attitude towards safety" is graded the second with a relative importance index of 0.89 and average index of 4.43. The results are a clear indication of poor worker's attitude towards safety. The respondent grade availability of safety equipment, safety inspections and organization safety policy are ranked the third, fourth and fifth respectively, with a relative importance index of 0.85, 0.84 and 0.82. As for the rest of the factors it is ranked the sixth, seventh, eighth, ninth and labour turnover rates are graded as least factor which affecting effective/improper safety management in construction projects with a relative importance index of 0.59 and average index of 2.87.

Table 1. Factors Affecting Effective / Improper Safety Management using Statistical Package Social Science (SPSS) Software

Factors Affecting Effective Safety Management in Construction Projects	Statistics			
	N	Mean	Std. Error	Std. Dev.
Organization safety policy	30	4.03	0.176	0.964
Safety meeting	30	3.57	0.157	0.858
Safety training and awareness	30	4.57	0.124	0.679
Availability of safety equipment	30	4.27	0.135	0.740
Safety inspections	30	4.23	0.164	0.898
Safety incentives and penalties	30	2.77	0.124	0.679
Worker's attitude towards safety	30	4.43	0.124	0.679
Labour turnover rates	30	2.97	0.155	0.850
Compliances with safety legislation	30	3.77	0.190	1.040
Safety measuring devices	30	3.10	0.211	1.155

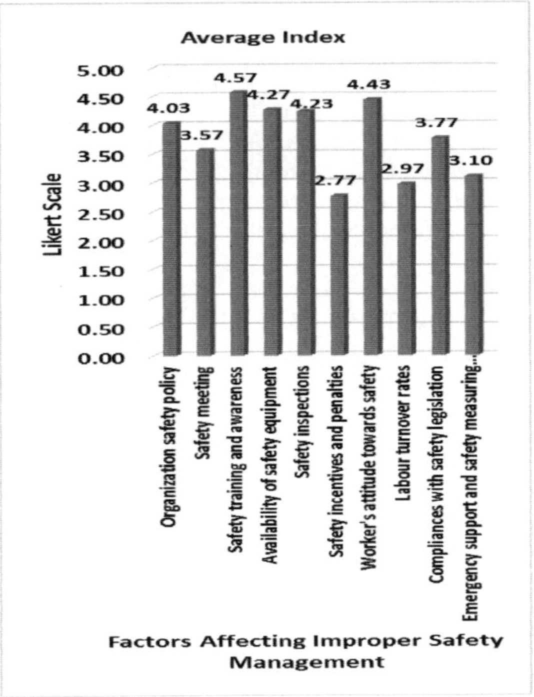

Figure 2. Factors Affecting Effective / Improper Safety Management Using Average Index

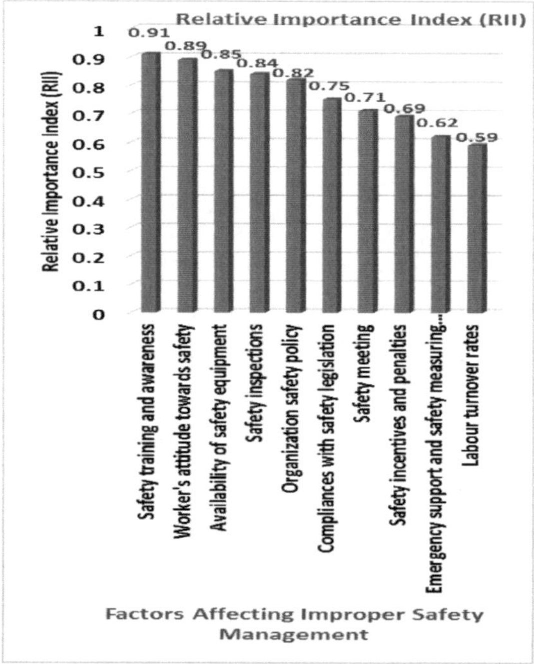

Figure 3. Factors Affecting Effective / Improper Safety Management Using RII

5. Conclusion

Construction worksites are one of the most high risk places because of the high safety cases reported at construction sites. This study investigates various ways that management may take in line to make their sites safer. This study analyzed all the variables that significantly will give effect to site safety. The results of this study are very important because the factors that affecting safety have been identified. Safety training and awareness, worker's attitude towards safety, availability of safety equipment, safety inspections and organization safety policy are the top five most important factors affecting safety at construction sites. If these factors are addressed and monitored closely, safety cases at construction sites can be minimized. In conclusion, safety affects all levels of the construction organization from government, company management, supervisor, and worker.

References

[1] Ahmadon Bakri et al (2006), Occupational Safety and Health (OSH) Management Systems : Towards Development of Safety and Health Culture. Proceedings of the 6th. Asia- Pacific Structural Engineering and Construction Conference (APSEC 2006), 5-6 September, 2006, Kuala Lumpur, Malaysia.

[2] Abdul Rahim Abdul Hamid and Muhd Zaimi Abd Majid (2006), Construction Safety Benchmarking, Proceedings of the International Conference In The Built Environment In The 21st Century (ICiBE 2006), 13-15 June, 2006, Kuala Lumpur.

[3] Othman, I., Napiah, M., Nuruddin, M. F., Klufallah, M. M. A., Effectiveness of preventive safety management in construction, Engineering Challenges for Sustainable Future - Proceedings of the 3rd International Conference on Civil, offshore and Environmental Engineering, ICCOEE 2016, 155-158.

[4] Abdul Rahim Abdul Hamid, Wan Zulkifli Wan Yusuf and Bachan Singh (2003), Hazards at Construction Sites. Proceedings of the 5th. Asia-Pacific Structural Engineering and Construction Conference (APSEC 2003), 26-28 August, 2003, Johor Bahru, Malaysia.

[5] C.Wu, F. Wang, P. X.W. Zou & D. Fang (2016). How safety leadership works among owners, contractors and subcontractors in construction projects, International Journal of Project Management 34, 789–805.
[6] Chang, L., editor. Preparing for construction in the 21st century, New York, ASCE, 97-102. Department Of Safety and Health, DOSH (2012)
[7] Edward J. Jaselski et al (1996), Strategies for Achieving Excellence in Construction Safety Performance. Journal of Construction Engineering and Management, March 1996, ASCE, USA.
[8] Muiruri, G & Mulinge, C (2014), Health and Safety Management on Construction Projects Sites.

Woman Friendly Mosque, Features and Facilities: A Case Study on Masjid Sultan Ibrahim, Universiti Tun Hussein Onn Malaysia

A Suratkon[1], N N Abd Salam[2], M H Rahmat[3], A S Mohd Arhan[4], I Abd Wahab[5], and S A. Ghaffar[6]

Department of Architecture and Design Engineering, Faculty of Civil and Environmental Engineering, Universiti Tun Hussein Onn Malaysia, Malaysia.

E-mail: nurnasuha@uthm.edu.my

Abstract. Mosque has gone through a long and complex development throughout the Muslim's history which involved the interrelation between its functions and its design. During the process, many aspects of its design were reconfigured including the female facilities in the mosque. In Malaysia, there is a progressive movement to transform the roles and functions of the mosque from exclusive place for prayers and seclusion to a community center. Nevertheless, there are still a numbers of muslim women voices their frustration towards the facilities provided to them in the mosque. Indirectly, this has discourage participation of women in the mosque. The objective of this paper is to investigate the opinions and needs of muslim women in a mosque and to propose appropriate features and facilities for a female-friendly mosque. The study explored through a field observation and a set of questionnaires, which involves a case study of Masjid Sultan Ibrahim in Universiti Tun Hussein Onn Malaysia. The field observation was carried out to analyze its current conditions, facilities, design and features. On the other hand, a set of questionnaire was distributed to the women visitors who were mainly students and staffs as to get their insight on this issue. From the field observation and the survey distributed, it was clearly seen that the facilities provided for the female users were still not up to the standard of a 'women-friendly mosque'. Some proposals were provided to improve on the current condition and to be taken into consideration by future designers in designing a mosque.

Keywords: Female friendly mosque, mosque design, Masjid Sultan Ibrahim

1. Introduction

A mosque or masjid (in Arabic) is a place where muslim prostrate and perform their obligatory prayer. The word masjid in Arabic comes from the verb 'sajada' which means to prostate [1]. In the earliest day of Islam, the roles and functions of the mosque is much wider than it is today. It was not only a place of prayer, but also the center for the Muslim community were religious instruction, political discussion and schools were indispensable appendage to the mosque [2]. When prophet Muhamad and his companion migrated to Madinah, one of the first thing that they did were establishing a mosque. Once the mosque is established, it became the center of Muslims activities for every level of community, being it a man, women or children [2].

1.1 The Development of Mosque's Functions and Its Design

Mosque as a center for Muslim community has become a tradition for centuries, since its foundation by Prophet Muhammad (pbuh) until Muslims civilization reached their peak of excellence during Fatimiyyah Era [3]. The idea of a mosque as the center for the community has been reflected architecturally in its design and its development is paralleled to the rapid expension of Islam. For example, during the Makkan and Madinan periods, there were no barriers between men and women

Content from this work may be used under the terms of the Creative Commons Attribution 3.0 licence. Any further distribution of this work must maintain attribution to the author(s) and the title of the work, journal citation and DOI.
Published under licence by IOP Publishing Ltd

praying area and no separate entrances which appear to support full female access to the mosque [4]. Furthermore, there was a designated space for children which was attached to the prophet mosque as to encourage children participation without compromising the sacred rules and courtesies of the mosque [5]. During Fatimiyyah era, mosque institutions such as the Al-Azhar mosque in Egypt becomes the education hub for the Muslims world by the 10th century AD [3]. An exemplary effort during this period is the idea of student accommodation in the mosque for students who are pursuing their studies which was known as Khan [3]. From these records, the early mosques in the Muslim world unswervingly retained its role as the center of Muslim life and supports full participation from every level of community [6].

The specialization of mosques from its role as the center of community to exclusive place for ritual prayer and seclusion has started during the 'Abbasid' period. The Abbasid period saw the establishment of Dar al Hakam as a special courthouses and *madrasas* as a formal educational institutions. On the other hand, the palace and governor's residence became the administrative and political centers [3]. These places stripped the role of the mosque as a holistic institution to an exclusive place for ritual prayers and seclusion only [6]. In Malaysia, the collapse of the Ottoman Empire in 1924 and the influence of secularism by British colonialism has further witnessed the reduction of roles and functions of the mosque. The British colonialism policy was to separate administrative and legal system of the states from Islamic belief [7]. Furthermore, the direct involvement of British education policy has caused the Islamic Education in *pondok, madrasas* and mosque to deteriorate gradually [8]. The role of religion becomes only limited to private matters and the mosque institution becomes exclusive to ritual and congregation prayers which is usually dominated by man. These policies might indirectly influenced the limitation of female excess thus results in the inadequacy of their facilities in the mosque.

Today, there has been a rise of Islamic understanding on the true roles of mosque in Muslim community. Islamic Centre and Human Development of Universiti Malaysia Pahang suggests that the mosque should be an institution that encourage the development and wellbeing of the Muslim community [9]. The Department Of Town And Village Planning Peninsular Malaysia, under the Ministry of Urban Wellbeing, Housing and Local Government also shows considerable support to this movement. They formulate a planning guide to ensure that the planning of mosque by the state, local and religious authority, developer and architect in Malaysia is in line with the aspiration to make the mosque as a center of Muslim life [10]. Architecturally, the planning guide focuses on two scopes which are the location of the mosque and also the space layout of the mosque. The space layout of the mosque can be divided into two parts, the external space such as green landscape, resting area and parking and the internal space compromising the *mihrab*, main prayer area, supporting facilities, the tower, ablution area and toilet [10].

The objective of this paper is to evaluate the current condition of one of the central mosque in Parit Raja, Batu pahat – Masjid Sultan Ibrahim, Universiti Tun Hussein Onn Malaysia, in regards to the comfort of female visitors to the mosque. The evaluation was done through a field observation towards the female facilities and by identifying female user's needs and expectations through a set of questionnaire. The findings from these evaluations were summarized and analysed in order to gain more understanding in the subject matter and hence a comprehensive proposal for appropriate features and facilities for a female-friendly mosque can be justified.

2. Field Observation

The purpose of the field observation is to evaluate the 'female friendliness' of the female facilities inside the mosque. The areas that were observed are the circulation area, women prayer area, women ablution area, and women waiting area. The findings from the observations are as follow:

The circulation area for the female visitors (i.e. route to the entrance, stairs connecting the prayer space to the ablution area and toilet) were considered 'women-unfriendly' as they were not fully-covered and resulted in distressing feeling to the women as their aurah might be seen by the public while passing through these areas (Figure 1).

Figure 1 – Non-fully enclosed circulation route and ablution room

The women praying areas are located on the first floor and mezzanine floor. On the first floor, the space is located at the back of the main praying area and segregated from the men's praying area by removable partitions (Figure 2). On the other hand, the prayer space on the mezzanine level is fully dedicated for the women (Figure 3) and can be considered as wide and comfortable space. However, due to its location on the highest level, it is inaccessible to people with difficulties, especially for those who have knee-problems which are typical to old women.

Figure 2 – Women's praying area

Figure 3 – Women's praying area on mezzanine level.

The women ablution room is located on the ground floor which is quite far from the main praying area. The ablution room is not fully-enclosed as it is open at the top and could be clearly seen by the people walking up the stairs next to the ablution area (Figure 1). It causes uneasiness feeling as again, the *aurah* of the female visitors will be exposed while doing the ablution. Apart from that, they had to immediately put on their hijab and unfold their shirts' sleeves right after finishing the ablution without having time to keep themselves dry from the ablution process.

The women waiting rooms are located on the ground floor which are visually and physically disconnected from the main praying area. One room is located in front of the women's ablution area (Figure 4) and the other is an enclosed room which also functions as Fardhu Ain Room (Figure 5). Both of the rooms were not well maintained and had defeated its main purposes. If there is an event occurring in the main praying area, people in those two spaces automatically disconnected and omitted from that event. These places ended up as abandoned places and less utilized by the visitors.

Figure 4 – Women's waiting room

Figure 5 – Fardhu Ain room

3. Questionnaires Distribution

A set of questionnaires were distributed to the women visitors to identify their needs and expectations. The survey took place for three weeks and the pilot study was conducted in the first week of the questionnaire distributions, specifically to test the competency and effectivity of the questions distributed. The questionnaire was consisted of four parts. Part A and Part B were focusing on the respondent's personal details and background while Part C and Part D were to find out the respondent's respond on the conditions, designs, features and facilities of the mosque. The survey was distributed on Monday and Thursday because the number of women visitors were the highest on those days. A total of 91 respondents have taken the survey.

The result from the survey had shown that 7% of respondents visited the UTHM's mosque every day, 54% twice a week, 22% monthly and 17% yearly (Figure 6). From this result, we can say that the women attendance to this mosque is quite low as only 7% of the respondent comes to the mosque every day. On the other hand, 98% of respondents do not have children and only 2% of respondents bring their children to the mosque and these respondents have experienced difficulties affiliate with children at the mosque.

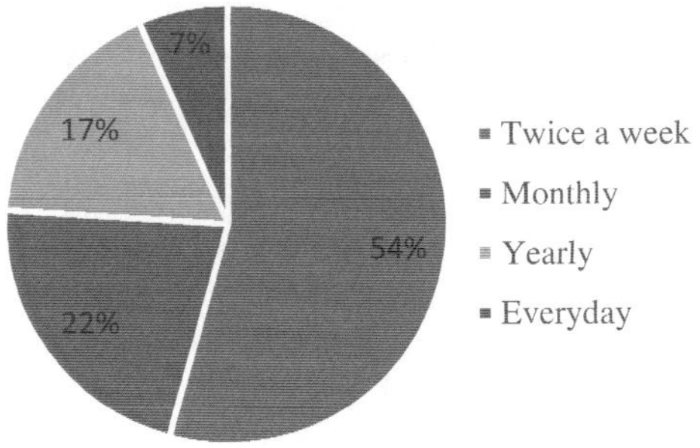

Figure 6: Percentage of the respondents coming to the UTHM Mosque

From the study, it has shown that there were some expectations to be fulfilled in order to make the mosque more women-friendly. 71% of the respondents requested for a children play/waiting area. 63% of the respondents suggested installing LCD screen in the women's praying area so that they could clearly see the person giving the speech in any event organised by the mosque's committee. Graph in Figure 7 and 8 summarised the key to understand women's needs and expectations:

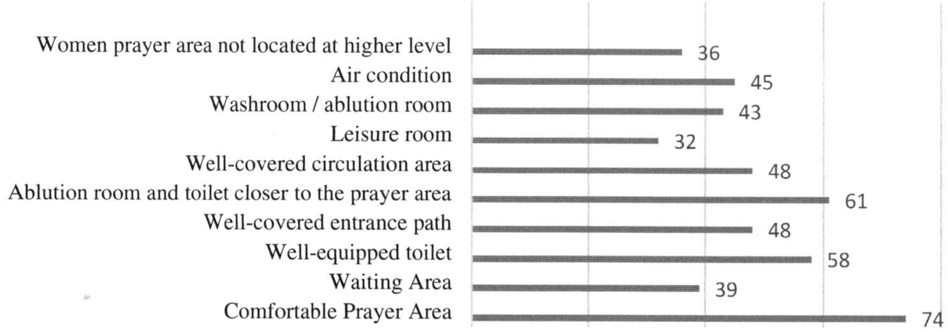

Figure 7: Perception of respondents towards new facilities for a female-friendly mosque (percentage).

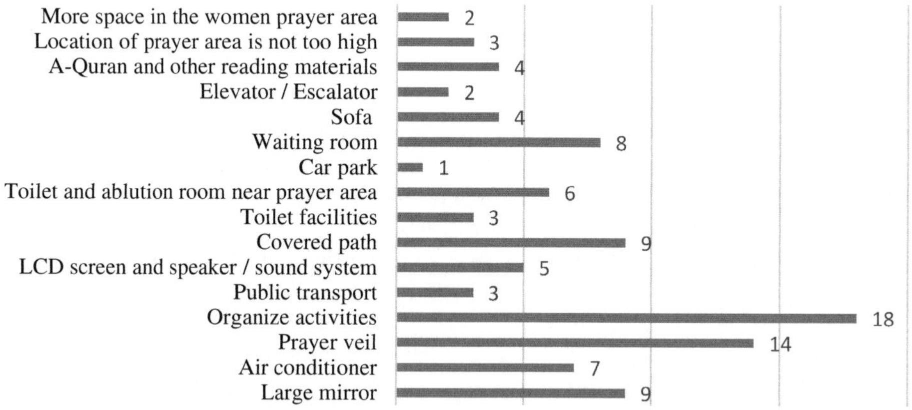

Figure 8: Number of respondents based on their personal opinion

4. Proposal for Improvement for Future Development

From the field observation and survey, suggestions for improvisation can be divided into two parts. The first part is the reconfiguration of spatial layout between the main prayer area and ablution room which ideally should be designed as one unit and attach to each other. These spaces should be fully enclosed, including the route connecting those two spaces. It is important to have a fully enclosed place because the process of doing ablution and getting ready for prayer for women involves uncovering and covering the veil. Furthermore, it is a practical strategy for architects to design an enclosed route between both spaces when the spaces are attach together.

Second is the additional and improvement of supporting facilities that fulfills the need of women and children. Well supporting facilities should be considered in the design of a mosque for the women to have comfortable experience while performing their obligations to the Almighty. One of the facility that should be considered is a well-equipped children area. Usually, the lecture taken place in the mosque takes almost 1 to 2 hours. For the mothers to have a comfortable learning process, a well-equipped children area near the main prayer area should be provided. On top of that, the women waiting area needs a lot of improvement. In Islam, woman who are in their monthly cycle are prohibited from entering the main prayer area but this is not the reason for them to be omitted from the mosque's activities. Installation of better speaker and sound system together with the LCD screen in the women prayer area and waiting area is highly recommended so that they could clearly see the *imam* or speaker during the prayers or any educational events.

5. Conclusion

This paper has elaborated on the requirements needed in implementing a women-friendly mosque. A women-friendly mosque is vital in encouraging women participation to the activities held in the mosque. From the study, four main areas have been identified as the area that needs attention and improvement as to become a female friendly mosque. Therefore, several architectural design proposal have been suggested in regards to those spaces. However, a comprehensive planning guideline should be formulated to guide the state, local, religious authority, developer and architect in designing a mosque that is friendly to women. Therefore, a study on a comprehensive planning guideline that includes female friendly features and design can be a topic of exploration for future research.

Acknowledgments

We thank the Office for Research, Innovation, Commercialization and Consultancy Management and the Registrar Office of Universiti Tun Hussein Onn Malaysia for supporting this paper and research.

References

[1] Abdulaziz A M., Mohamed A K and Nidzamuddin Z 2008 *Fungsi dan Peranan Masjid dalam Masyarakat Hadhari: Kepentingan Dan Peranan Masjid Dari Sudut Pendidikan Dalam Masyarakat Islam* (Johor: Universiti Teknologi Malaysia Press) p 61-79

[2] Zaimeche S 2002 *Education in Islam: The Role of the Mosque* (United Kingdom: Foundation for Science Technology and Civilization) p 3-6

[3] Haron Z, Ripin M N and Yusof F M 2008 *Fungsi dan Peranan Masjid dalam Masyarakat Hadhari: Masjid Sebagai Institusi Pendidikan, Sorotan Abad Ke-3 H - 5 H* (Johor: Universiti Teknologi Malaysia Press) p 2-20

[4] Reda N Women in the Mosque: Historical Perspectives on Segregation *American J. Islamic Soci. Scie.)* **21** 78-97

[5] Mat Said A H A C H 2008 *Fungsi dan Peranan Masjid dalam Masyarakat Hadhari: Peranan Masjid Dalam Masyarakat Hadhari* (Johor: Universiti Teknologi Malaysia Press) p 29-45

[6] Rasdi M T M 2007 Mosque Architecture in Malaysia: Classification of Styles and Possible Influence *J. Alam Bina* 37

[7] Aziz A and Shamsul A B 2004 The religious, the Plural, the Secular and the Modern: a Brief Critical Survey on Islam in Malaysia *J. Inter-Asia Cult. Stud.* **5** 341-56

[8] Harper T N 1999 *The End of Empire and the Making of Malaya* (Cambridge: Cambridge University Press)

[9] PIMPIN 2017 *Kepentingan Masjid Memperkasa Legasi Ummah* (Pahang: Universiti Malaysia Pahang)

[10] Desa P B D 2011 *Garis Panduan Perancangan Masjid Dan Surau* (Malaysia: Jabatan Perancangan Bandar dan Desa) p 1-2

Challenges in the Implementation of Green Home Development in Malaysia: Perspective of Developers

Rumaizah Mohd Nordin[1], Ahmad Hafizi Abd Halim[2], Julitta Yunus[3]
Faculty of Architecture, Planning & Surveying, Universiti Teknologi MARA, Shah Alam, Malaysia[1,2,3]

drrumaizah@salam.uitm.edu.my[1]

Abstract. The construction industry is the main contributor to Malaysian economic growth. One of the main sectors to be highlighted by the Government is housing sector. The population in Malaysia is increasing year by year, thus the demand for houses also rises significantly. However, the fulfilment of basic needs for people normally gives the adverse impacts to the environment. Green home concept is introduced with a purpose to create an acceptable standard of living as well as to preserve nature from destruction, whilst to promote efficiency of energy, water and other natural resources. The study aims to identify current level of awareness and understanding regarding the Green Home concept, together with identifying the challenges that contribute to lack of initiatives in implementation of green home development. This study utilizes the qualitative methodology utilising interviews with housing developers operating in Klang Valley. 20 respondents were interviewed with a semi-structured interview. This study found that level of awareness and understanding on green home concept among construction players and public is low to moderate level. However, it shows improvement in terms of its implementation with all respondents agreed that cost factor is greatest challenge to its implementation. Other challenges identified from the study are low awareness and understanding among construction players and public, low demand for green home, and lack of Government enforcement and initiatives. The study is eventually intended to enhance and improve the sustainable practice in Malaysian's construction industry.

1. Introduction

Malaysia plans to reach the status of a developed country in the year 2020 and the construction industry is regarded as the main contributor to achieve this vision [1]. Housing sector contributes to Malaysian economic growth, as the industry provides job opportunities for 800,000 people in Malaysia, which represents 8% of total workforce [2]. Adding to that, housing can be inferred as an industry with activities that provide buildings and structures with the aim to fulfil the human needs for shelter, protection from any dangers and as a rest pace after daily activities [3]. Besides, the population in Malaysia keeps growing year by year leading for the increasing demand of housing development [4]. Other than that, the housing development should be complement with other basic amenities such as electricity, water supply, recreational facilities, medical services, community centres and so forth towards improving welfare of people [1]. However, the urban expansion will take place at green-field site and hillside as well as the reclamation areas which leads to water pollution, coastal erosion, wetland degradation, ecosystem disruption and fisheries problems [5]. As for that matter, the fulfilment of the needs of new housing development normally will give the impacts on nature such as the environmental degradation, scarcity of natural resources, land degradation, air and water pollution [6]; [7]; [8].

Content from this work may be used under the terms of the Creative Commons Attribution 3.0 licence. Any further distribution of this work must maintain attribution to the author(s) and the title of the work, journal citation and DOI.
Published under licence by IOP Publishing Ltd

In fact, the increasing effort in sustainable construction or green construction by construction industry is the crucial move in achieving sustainable development through environmental, economic and social issues [9]. This is true as green construction is aiming to create facilities and building that are sustainable with huge attention focuses on the satisfaction of the end user [10]. Hence, green housing built to provide house that uses less energy, water and natural resources, reduce waste and provide healthier indoor environment compared to conventional house [5]. Besides, green housing can be defined as housing which is built with intention to fulfill needs of the current people without sacrificing the ability of future generation to meet their needs with least possible impact on the environment [1].

Despite the importance of green housing is pressingly, the level of awareness and knowledge on sustainable development among Malaysia's construction players is below average level [11]; [12]. Meanwhile, the level of implementation of sustainable development is within moderate level [5]. According to Abidin [13], the construction players are still trying to understand about benefits of sustainable practice. Besides, the sustainable housing development concept is still quite new in Malaysia and only few housing projects in Malaysia had applied the concept [11]. The main barriers of this are related to cost factor; financial constraint, risk of investment and lack of credit resources to cover up front cost [13]; [4]. Sustainable practices are believed to increase project cost which leads to higher selling price [8]. Moreover, the majority of Malaysian still prefer to buy conventional home than green home since it is cheaper [14]. The other barriers affecting the implementation of green home concept is low level of education, knowledge and experience on green design and technology [5]; [15]; [12]; [16]. Most of Malaysian developers are profit-driven, so they are reluctant to perform sustainability in their projects especially small developers due to low demand of high-priced green home from potential buyers [17]. Thus, it is noteworthy to identify the challenges and barriers that contribute to lack of initiatives in implementation of green home development in Malaysia.

2. Methodology

This study was conducted using qualitative approach, utilising semi-structured interview. Sampling frame for this study is housing developers operated in Klang Valley and out of 65 developers identified, 20 agreed to participate which representing 30.8 percent response rate. The respondents consist of top management level including Managing Directors, Landscape Architect, Managers and Executives.

3. Findings

Table 1 illustrates the respondents' profile including their years of experience, current position, and their appropriateness to be respondents of the interview session.

Table 1. Respondents' profile.

Respondents	**Appropriateness as Interviewees**		
	Involvement in GBI Project	**Having Authority to Respond**	**Years of Experience**
Managing Director (R5, R7, R13)		√ (R5, R7, R13)	14, 10, 10
Landscape Architect (R9)	√ (R9)		8
Manager (R2, R6, R8, R11, R14, R18, R20)	√ (R14, R20)	√ (R2, R6, R8, R11, R18)	10, 8, 10, 11, 8, 10, 12
Executive (R1, R3, R4, R10, R12, R15, R16, R17, R19)	√ (R1, R3, R10, R19)	√ (R4, R12, R15, R16, R17)	4, 9, 9,12, 8, 12, 10, 8
TOTAL	**7/20 (35%)**	**13/20 (65%)**	

In overall, 13 respondents (65%) have position in top management level, indicates that they have the authority to give their opinion and statement on the issues of green home development. Apart from that, half of them (50%) possess at least 10 years of experience in the construction industry. While, there were 7 respondents (35%) who represent their company to participate in the interview due to their involvement and experienced in handling green projects for their current companies. Furthermore, many of them are familiar with the assessment of GBI rating tools. This implies that the respondents have reasonable knowledge on the issue of sustainable construction and Green Home development.

Level of Awareness and Understanding on Green Home Development among Organizations.
Table 2 shows the level of awareness and knowledge on green home concept among 20 respondents that have been interviewed. In overall, 85% of the respondents (17 respondents) confirmed the level of awareness and knowledge among employees in their organization is low. Only few employees who involved in projects incorporated with green design/features have knowledge and aware about green practice. For instance, R1 revealed that the level of awareness and knowledge among staff is below moderate level. Only those who work in the Sustainability and Quality Management Department possess some knowledge on green home concept. Nonetheless, the company is moving towards enhancing awareness among its staff on the sustainability practice. As R1 put forward, *"The Company is committed to transforming spaces into sustainable communities where neighbourhood last for generations. To live sustainably, we create communities that are socially inclusive, economically vibrant, save and environmentally enriching. Sustainability is a mind-set that actively cultivates to deliver on these promises."*

Table 2. Level of awareness and knowledge among employees within respondent's organizations.

Level of Awareness and Understanding on Green Home Development among Respondent's Organizations.		
LOW	MODERATE	HIGH
17 (85%)	3 (15%)	0 (0%)

Whereas, the R2 revealed there are approximately 10% from the 80 employees in this company are aware and have knowledge on green home concept, which are working in Planning and Development Department. But then, according to R4 only 30% of their staff know and aware about green practice. They are mainly working within the Product Development and Implementation Team who knows little on theory and technical part of green design and technology. Other than that, the expertise engaged to advice on that matter. The Project Department Team is responsible only to administer the consultant's proposal and bring up to top management for approval, meanwhile the Implementation Team is responsible to supervise work during construction. The rest of the respondents raised their opinion of only staff (i.e., Architect and Engineer) who involve directly with design matters and project implementation possess average level of knowledge and aware about green home concept. They tend to engage external expertise to advice on green design and technology. In other situation, R7 confirmed his company involve solely in affordable housing construction, thus none of the employees are aware and have no knowledge on green home concept.

However, there are three (3) respondents (15%) who believe the level of awareness and knowledge among their staff are moderate. Likewise, R18 stated that most of the staff are not only aware on green building concept but also have knowledge on theory and technical skills on green design and technology. They also know the assessment on GBI rating works. On top of that, the top management level are committed to send their staff to technical courses/talks regarding green design and technology even it was in overseas. As for R19, the high level of awareness and knowledge among staff due to continuous initiative and effort done by the management to instil sense and knowledge on sustainability. Furthermore, the management also encourage their staff to learn and join courses on green design and technology. Similarly, R20 levelled the green awareness as moderate due to staff understanding is towards preserving the environment. Generally, the current level of awareness and understanding of developers' staff based on interview session in line with previous studies explained in literature review that it is within low to moderate level.

Challenges in the Implementation of Green Home Development
Table 3 presents the seven (7) challenges in the implementation of green home development clarified by all 20 respondents in the interview session. The cost factor is the greatest barrier that leads to low

level of the implementation of green home development. All respondents (100%) stated that cost is the significant issue that hinder them to practice green in their projects. They said that the green home development will increase construction cost. For example, R1 stated that green home development is very costly especially in material and system used. Solar panel and volatile organic compounds (VOC) paint identified as most costly items.

Table 3. Challenges in the implementation of green home development.

Challenges and Barriers	**Frequency n = 20**	**Percentage %**	**Ranking**
High Initial Cost **(H.C.)**	20	100%	1
Lack of Awareness and Knowledge among Developers **(D.A.)**	17	85%	2
Lack of Awareness among Public **(P.A.)**	15	75%	3
Lack of Awareness and Knowledge among Project Team and Contractors **(T.A.)**	14	70%	4
Low Demand from Potential Buyers **(L.D.)**	13	65%	5
Low Availability of Green Building Materials and Technology Locally **(G.M.)**	9	45%	6
Lack of Enforcement and Incentives from Government **(G.E.)**	4	20%	7

Other higher cost involved is GBI assessment, design fees and professional fees. This statement is supported by R9 and R11 whos emphasised on sustainable option normally requires for green rating assessment. According to R2, the higher level of green certification means the higher cost to be incurred (up to 30%) for mechanical and electrical equipment installed to achieve energy and water efficiency. Despite of having reduced operating cost with green home concept, the saving gained is not worth for the company. As mentioned by R3, R6, R8, R10, R14, R19 and R20. These findings are in line with previous researches including the green option normally requires higher capital upfront [13]; [18]; [19]. In addition, the extra expenses went for appointing environmental consultants, green rating assessment fees and procurement of new technology and green materials [5]. Following High Initial Cost (H.C), 85% of the respondents revealed that the low level of awareness and knowledge among employees in their organization is due to majority of site workers are not competent in sustainable practice. While, 3 respondents (15%) claimed that the level of ther staff on green awareness and knowledge is moderate. The result inline with the studies carried out by Said et al. [11], Aliagha et al. [12] and Algburi et al. [2]. Lack of Public Awareness (P.A.) ranked as third for challenges in the implementation of green home development by 75% respondents. They rated the level of public green awareness and knowledge as low. According to R1, majority of Malaysian people are not aware on sustainability issues, hence they are not ready to accept that concept as their lifestyle. Education also paly an important role. R10 claimed that the mentality of Malaysian must be changed first. For example, the ignorance attitudes towards switching off the electrical appliances when leaving the room. The green home concept will not achieve its target if the mentality and awareness level of public is still low. This findings inline with previous studies that explained about public awareness on green home concept [13]; [17]; [14]; [20]; [4]. Followed by Lack of Awareness and Knowledge among Project Team and Contractors (T.A.) is ranked fourth with 70% of the respondents agreed to this. The respondents viewed that current construction industry is lacking on expert professional and consultant in green design. They also highlighted that majority of the contractors are lacking in green construction knowledge and skill, which making them incapable to handle the job. Others, majority of project managers and contractors also are inexperienced handling the green building projects. This is due to the implementation of sustainable construction is still at pioneer stage in Malaysia. In overall, the result from the analysis shows similarity with findings gathered in the literature review regarding issues of lack of knowledge, and expertise among project team members and contractors on green design and technology [12]; [20]; [2].

Low Demand from Potential Buyers (L.D.) was ranked fifth (65%) as demand for green home depends on the selling price. Since, majority of Malaysian are low-middle income people, the green home price is beyond their affordable ranges. This situation is similar with Alias et al. [14] and Mohamed et al. [4] findings. Besides, lack of demand for green home is due to people's resistance to change their lifestyle also highlighted and agreed by Aliagha et al. [12]. Finally, Low Availability of Green Building Materials and Technology Locally (G.M.) is ranked sixth by the respondents (45%) as most of the sustainable materials, products and systems are not available here in Malaysia and requires to be imported from overseas (R3, R9, R10 and R15). On the other hand, most respondents believed that the green building materials and technology easily procured in the local market, however its rather very expensive. These are contradict with Alias et al. [14], Shari & Soebarto [21], Tey et al. [9] and Algburi et al. [2] which claimed that green building materials and green technology are limited in local market. Lastly, lack of Enforcement and Incentives from Government (G.E.) was found to be the least challenges that lead to low level of the implementation of green home development.

4. Conclusion

20 out of 65 housing developers operated in Klang Valley have been interviewed, which representing 30.8 percent response rate. The targeted respondents in this study include senior management levels in the organization; with experience and currently handling the green projects. The result shows the awareness among developers in Malaysia is still at low-moderate level. The study indicates the greatest barrier of implementing the green development is a higher of capital upfront i.e. professional fees, green assessment fees, design and material cost. The unavailability of financial incentives from the Government also added to the obstacle. Then, amongst the biggest challenge of the developers is to produce right yet affordable price of green home to attract low-medium income people. Other than that the obstacles pointed to improbable experiences and in-depth knowledge among the team members and low mentality of Malaysian towards sustainability concept. However, the study revealed there is some improvement in green materials availability in the market. Where it shows that the green building materials are easily procured locally unlike in previous studies.

The Government shall not only increase the green campaign but also improves their enforcement on regulation and guidelines on green home development. The effort on making green practice as a pre-requisite for the developers to secure their future projects can be a good start. Finally, the Government should highly reconsider to provide financial incentives to construction players as booster elements in increasing the green practices. This study recommends to future of green home study shall extent findings on the mechanism to solve the issue of reducing cost to improve the implementation of green home development in Malaysia.

5. References

[1] M. Ubale, D. Martin and S. T. Wee 2015 Investigating Housing Affordability Pursuant to Sustainable Development Mechanism and the New Malaysian Housing Policy *Journal of Economics and Sustainable Development* 49-66

[2] S. M. Algburi, A. A. Faieza and B. T. H. T. Baharudin 2016 Review of Green Building Index in Malaysia; Existing Work and Challenges *International Journal of Applied Engineering Research* **11** 3160-67

[3] F. A. Ibrahim, M. W. Mohd Shafiei, R. Ismail and I. Said 2014 Green Homes Development: Factors Affecting Housing Developers' Readiness *ARPN Journal of Engineering and Applied Sciences* **9** 971-980

[4] S. Mohamed, S. Y. Chau and M. H. I. Rahim 2015 A Review of Sustainable Development for Housing Scheme in Malaysia *Global Journal of Engineering Science and Research Management* 67-72

[5] N. Z. Abidin, N. Yusof and H. Awang 2012 A Foresight into Green Housing Industry in Malaysia *World Academy of Science, Engineering and Technology: International Journal of Environmental, Chemical, Ecological, Geological and Geophysical Engineering,* **6** 55-63

[6] F. Ibrahim, M. Shafiei and R. Ismail 2013 Malaysian Housing Developers' Readiness in Green Homes Development *World Applied Science Journa* **28** 343-52

[7] W. M. N. M. Fuad and T. H. T. Ismail 2012 Examination of Green Building Material (GBM) Usage in Construction Among Malaysia's Local Professionals as a Means for Sustainable Development *Environmental Research Project* 1-16

[8] Y. H. Chan, B. C. T. Lee and J. C. Lee 2014 Sustainability in the Construction Industry in Malaysia: The Challenges and Breakthroughs *World Academy of Science, Engineering and Technology: International Journal of Social, Behavioural, educational, Economic and management Engineering* **8** 1209-13

[9] J. S. Tey, K. C. Goh, T. W. Seow and H. H. Goh 2014 Challenges in adopting sustainable materials in Malaysian construction industry *eprints.uthm.edu.my/4081/*

[10] A. H. A. Thanim and M. H. Abdul Samad 2016 A Review on Sustainable Design and Indoor Thermal Comfort of a Green Building *ARPN Journal of Engineering and Applied Siences* **1** 3712-17

[11] I. Said, O. Osman, M. W. Mohd Shafiei and A. Razak 2009 Sustainability in the Housing Development Among Construction Industry Players in Malaysia *The Journal of Global Business Management* **5** 1-9

[12] G. U. Aliagha, M. Hashim, A. O. Sanni and K. N. Ali 2013 Review of Green Builing Demand factors for Malaysia *Journal og Energy Technologies and Policy: Special Issue for International Conference on Energy, Environment and Sustainable Economy (EESE 2013)* 471-78

[13] N. Z. Abidin 2010 Sustainable Construction in Malaysia - Developers' Awareness *World Academy of Science, Engineering and Technology: International Journal of Social, Behavioral, Educational, Economic, Business and Industrial Engineering* **3** 480-87

[14] A. Alias, T. K. Sin and W. N. A. W. A. Aziz 2010 The Green Home Concept - Acceptability and Development Problems *Journal of Building Performance* 130-39

[15] K. C. Goh, S. Tawee and H. H. Goh 2013 Challenges of implementing sustainability in Malaysian housing industry *eprints.uthm.edu.my*

[16] N. Zainordin and C. T. Y. Mei 2015 An Insight of Sustainable Development - A Study Among Construction Professional in Malaysia *International Journal of Scientific & Engineering Research* **6** 45-50

[17] N. Z. Abidin 2010 Investigating the awareness and application of sustainable construction concept by Malaysian developers *Habitat International* **34** 421-46

[18] S. V. Lim 2011 The Use of Green Building Materials in The Construction Industry In Malaysia *PhD thesis, Universiti Tunku Abdul Rahman (UTAR)*

[19] M. Samari, N. Ghodrati, R. Esmaeilifar, P. Olfat and M. W. M. Shafiei 2013 The Investigation of the Barriers in Developing Green Building in Malaysia *Modern Applied Science* **7** 1-10

[20] M. Shafiei, M. Samari and N. Ghodrati 2013 Strategic Approach to Green Home Development in Malaysia - the Perspective of Potential Green Home Buyer *Life Science Journal* 3213-21

[21] Z. Shari and V. Soebato 2012 Delivering Sustainable Building Strategies in Malaysia Stakeholders'Barriers & Aspirations *Alam Cipta Universiti Putra Malaysia* 3-12

Tracing the Influence of Stakeholders of the Nusantara Masjid Architecture

Izzat Asyraf Mahyuddin[1], Zaiton Abdul Rahim[2], Norwina Mohd Nawawi[2]

[1,2]Kuliyyah or Architecture and Environmental Design, International Islamic University Malaysia, Kuala Lumpur

Abstract. Masjids or mosques were built across the Malay Archipelago or the *Nusantara*, as a statement on the rise and strengths of the Islamic community since the arrival of Islam in the 1300s. As the latest religion to the South-East Asia after Hindu-Buddhist, the mass often discusses the form and make of the masjid but seldom question who's who constitutes the design and the rigmarole of masjids. This paper aim at identifying the stakeholders of masjid architecture of Nusantara through analysing the involvement of stakeholders from historical and primary data, covering the process of initiation, through masjid construction including the conceptions and perceptions on the factors influencing the architecture of the masjid. The paper intents to demystify the sole responsibility of the architect in designing masjid of Nusantara region and places the will of Islam as the religion of initiation and stakeholders, whom are the rulers, scholars and common Muslims, as the *Ummah*, the reason of its make and form it is today.

1. Introduction

A masjid is a place where Muslims gather to perform daily prayers, one of which is an obligation imposed on every Muslims five times a day. The word masjid itself derived from the Arabic word "*sajd*", meaning to perform worships and refer to the act of "*sujud*", an act of prostrating on the ground carried out during prayers. A masjid is a sanctuary, considered as the best of places by Prophet Muhammad p.b.u.h., in which the rewards of congregational prayers increase up to 27 times than praying by oneself. Masjid is the centre of Muslim community where many religious activities take place, weekly Friday congregational prayer, yearly congregational prayers marking the end of fasting month (Eidul Fitr) and pilgrimage (Eidul Adha), sacrificial of animal during Eidul Adha (Qurban), religious classes, solemnization of weddings, and other religious activities.

History indicates that Islam in the Malay Archipelago was spread through missionaries who came to the Malay world. The religion was accepted through peaceful and voluntary conversion (Arnold, 1913). Propagation of Islam was largely credited to the works and efforts of the Arab and Indian merchants (Mahmud, 1960). However, there is no concrete evidence on the exactly date of the first introduction of Islam in the Malay world. Although contacts with Muslim merchants are reported as early as 7th century, it was not until the 13th century that Islam gained a significant influence (Mohd. Suhaimi, 2014). Islam in the Malay world was not apparent until it was accepted by rulers of the Malay Archipelago, resulted in the spread of the religion throughout the local community. Islam spread though missionaries from Arab and Indian who settled down in regions of Nusantara, not only to ensure the propagation of the

Content from this work may be used under the terms of the Creative Commons Attribution 3.0 licence. Any further distribution of this work must maintain attribution to the author(s) and the title of the work, journal citation and DOI.
Published under licence by IOP Publishing Ltd

religion but to be a point of reference for the local community. Islamic schools in the form of *madrasah, pasantren* and *pondok* were built to teach teachings of Islam and eventually produced local scholars. The many masjids around the Malay Archipelago were testament of the spread of Islam in the Malay world.

Prophet Muhammad (p.b.u.h) said, '*whoever builds a masjid will be heftily rewarded in the Hereafter*'. This hadith reflects the importance of masjids and many people are encouraged to build masjid. However, it is unclear how the construction of masjid in Nusantara comes about, i.e. who are the stakeholders responsible for the construction and architecture of the masjids? What are the cultural and religious values that influenced it? Is the masjid architecture influenced by the stakeholders and in what ways? Lee (2008) defines stakeholders as those in the community who only have interest in what is occurring. This sets apart on who is involved in the process and who is not. This allows for a more comprehensive approach in any decision making. Lee definition is consistent with that by Friedman (2006), Aoki (1984) and Fontaine, Haarman and Schmid (2006). A stakeholder is either a person, group or organization that has interest or concerns in an organization (Business Dictionary, 2017). These variety of people can have an effect in the decision-making policies and having a say on the direction of intent for a cause. For the purpose of this paper, stakeholders is defined as the person (s) who initiated, built or involve in the decision to build and construct a masjid.

Review of literature indicates the earlier masjids reflect the Nusantara architectural styles. However, external influence was evident from some of the Nusantara masjids. Pre-Islamic Hindu architecture was also prevalent in the transition period when the Malays came to accept Islam partly due to the fact that the Islamic scholars did not undo or immediately changed the local culture but gradually change it consistent with the level of acceptance of Islam as a way of life. Nusantara masjids built always includes elements of the regions with its traditional values, due to the strong attachment to the traditional local culture (M. Kassim, 2014). Harun (2005), reasoned that since the widespread of Islam in the early 14th century, majority of the community in Nusantara embraced Islam and it completely changed their lifestyles.

The architecture of the masjid in Nusantara differ between regions, but retained similarities such as the roof form, shape, materials, raised on platform and square in shape. Differences can be seen in the specifics of the masjid such as the ornaments that have local cultural influence. Yet some of the masjids have design elements that were alien to the local culture such as seen in Masjid Al-Imtizaj in Bandung, Masjid Cheng Ho, Surabaya; both in Indonesia and Masjid Kapitan Keling, Penang and Masjid Tengkera, Malacca in Malaysia. The architecture of these masjids appear to foreign elements yet accepted and embraced by the locals. The variation in terms of design reflects that the architecture of the masjids may be influenced by those responsible in building it. Based on the discussion above, the objectives of the paper are to examine the role of stakeholders in masjid Nusantara: To identify the stakeholders of masjid Nusantara architecture in Indonesia and Malaysia and to examine the specifics of masjid Nusantara architecture influenced by the stakeholder

2. Methodology

Recorded information and resources concerning stakeholders of the traditional masjid in the Malay Archipelago is not conclusive. This study is at its early inception is exploratory. It employs secondary data from past studies on traditional masjids, field work and observation and interviews whenever relevant. Specifically, data on masjids in different locations in Indonesia comes from journal articles that presented detail discussion on a particular masjid, measure drawing and reports from Heritage Studies, International Islamic University Malaysia, fieldworks and observation. Data on masjids in Malaysia were gathered from fieldwork, observation supported by information from past studies. Seven masjids (3 from Indonesia and 4 from Malaysia) were as shown in 1 Table 1.

IOP Conf. Series: Materials Science and Engineering **291** (2017) 012021 doi:10.1088/1757-899X/291/1/012021

Table 1: Case and Source of data

Name	**Location**	**Date of Construction**
Masjid Lima Kaum, Batu Sangkar	West Sumatera, Indonesia	1650
Masjid Jami' Sulthan Ayyub Sanggau,	West Kalimantan, Indonesia	1826
Masjid Agong Sumenep	Sumenep, Madura	1779
Masjid Kampung Laut	Kelantan, Malaysia	17TH Century
Masjid Kampung Tuan	Terengganu, Malaysia	1830
Masjid Kampung Hulu	Melaka, Malaysia	1720
Masjid Tengkera	Melaka, Malaysia	1728

Table 2: Method of data collection

Name	Journal articles	Measured drawing	Fieldwork
Masjid Lima Kaum, Batu Sangkar	•		•
Masjid Jami' Sulthan Ayyub Sanggau,	•		
Masjid Agong Sumenep		•	
Masjid Kampung Laut			•
Masjid Kampung Tuan	•		•

Masjid Kampung Hulu	•		•
Masjid Tengkera	•		•

3. Findings and Discussions

Review of literature indicates different stakeholders involved in building the old masjid in the Malay Archipelago or Nusantara namely the scholars who propagated Islam in Nusantara, the rulers who came to accept Islam and eventually built masjids to spread the teaching of Islam and Muslim traders or individuals who came to the shore of the Malay Archipelago. The person who built, oversee or responsible for the construction of the masjid have significant influence on the specifics of the architecture. The following discussions elaborate on the role and influence of stakeholders in the old masjids of Nusantara. Regardless of the stakeholders, Islamic teachings were symbolically translated into the design of the masjid such as the number of columns or roof tiers in relation to the pillars of Faith in Islam, the greatness of the Creator and other teachings, etc.

The widespread of Islam developed through systematic teachings in traditional religious school (*madrasa, pesantren* and *pondok*), resulted in scholars and preachers who would continue spreading the religion to the people. The need to build a masjid is not only for prayers but also to introduce and teach Islam to the people. In the history of Islam in Indonesia, the Islamic scholar such as Wali Songo did not immediately change the social or changed the local cultural values that has developed over a period of hundreds of years with newly introduced Islamic values. This is because Islam was very different from the belief people were used to. Efforts were made to introduce Islamic teachings through the construction of masjids where newly converts could learn about the new religion and seek refuge if needed. The masjid started with a humble beginning; a simple open structure raised on stilt or on a platform. This small shed evolved into small establishments for studying the teaching of Islam to the community, commonly called *pesantren* (Mastuhu, 1994). It is believed that this is the starting point of how masjids were later introduced into the community. Dome and minarets were not used in the architecture of the masjids but instead the masjids benefited from exiting Hindu religious building. The architectural styles of masjids emerged gradually, reflecting important changes consistent with the political situation and technology of the people (M. Kassim, 2014).

Zainuddin (2013) discusses the many stakeholders involved in the conception and inception of Masjid Lima Kaum, Batu Sangkar, one of the oldest masjid in Sumatra Barat who includes the scholars, community leaders and members, and the builder who supervised its construction. The involvement of the stakeholders influences the masjid architecture in different ways; the five tiered roofs and the entrance roofs represented the people in the five *kaum* or villages in the area who worked together to construct the masjid, the 11 main pillars represents the scholars and the minimal carvings and ornament represent the culture of the people. Another Nusantara masjid, Masjid Tuo Kayu Jao has 11 pillars, representing the 11 Muslim scholars who were responsible in its construction (Zainuddin, 2013). The number 11 has been a symbol in the Minangkabau custom, and whenever Sheikh Burhanuddin Ulkan, the scholar responsible in propagating Islam in West Sumatra, preached in an area, he would send 11 scholars to build a masjid and this was reflected in the main pillars of the masjid (Zainuddin, 2013) as seen in many masjids in West Sumatra.

The masjid is simple in design and have minimal ornamentation and embellishment. The masjid Kampung Laut in Kelantan, Malaysia was built in the 17th century by an Islamic scholar who was on the way from Pattaya, Thailand who introduced and spread Islam to the local community. The masjid was originally a simple timber structure, 400sqft in area, with only 4 columns to support a three-tiered roof made out of palm leaves. It underwent multiple renovations with addition of new spaces and walls, 20 columns and auxiliary spaces such as the *serambi*, additional prayer spaces and a minaret. The masjid was simple in design and has very minimal ornamentation and devoid of any embellishment and remained to maintain its style until today.

Masjid Kampung Tuan, Terengganu was built by Syed Abdul Rahman Abdul Samad, an Islamic scholar who was from Pattani, Thailand who fled the country in 1830 due to the on-going war and found his way to preach Islam to the local Malays. He built a *musolla* as the number of people attending his preaches increases. The *musolla,* based on the characteristics of Siam architecture is 9.14m x 9.14m uses a single *cengal* tree, simple in design, similar to Masjid Kampung Laut. Situated next to the river (Sungai Limbong), it has four-tiered roof, a replica of *Buah Butung* made out of copper was installed on the roof, originally meant for Masjid Keresek in Thailand. The whole masjid was built by using a single *cengal* tree located in the area. The openings in the masjid was limited to 60 cm wide windows with no doors to give spiritual concentration to the patrons. There are 12 columns on the outer perimeter of the structure, connected without the use of nails. There is no information on the reasons behind the number of columns or other architectural features of the masjid.

Priyono (1991) noted that the history of Islam throughout Nusantara can be identified since the 11th century, but the development of the widespread of Islam came in the early 14th century, when the central indigenous authoritative power, Majapahit was declining. It is not until the early 16th century that the establishment of Demak maritime state as the first Islamic Kingdom in Java (Sulasman, 2013) came into proceedings. Demak then started to expand its authority through various means of both military and economy conquests throughout Java. The intent was to integrate religion, economy and political power. Many areas accepted Islam and under the rule of the new sultanate. This later on spurred the rise of later Islamic kingdoms rules my Muslim rulers such as Banten and Cirebon. The influence of the Muslim rulers on the masjid architecture can be clearly seen with the construction of masjid within the premise of the palaces to indicate the importance of Islam.

The influence of Demak can also be seen in the masjids in Nusantara regions which follow the same style with a mixture of the local architecture and the teaching of Islam as seen in the Masjid Jami' Sulthan Ayyub Sanggau, West Kalimantan. M. Kassim (2014) in his analysis found the architecture of the masjid a blend of Islamic architecture with Hindu-Javanese influence. Islamic architectural elements can be seen from the calligraphy that adorns almost every side of the ceiling of the mosque. Green colors of the masjid gives a peaceful ambience and yellow symbolizes the kingdom as the masjid was built by a royal, Sultan Ayyub, one of the princes of the missionary who introduced Islam to the area, married the local leader and became the first ruler in the area. The masjid was built in 1826, located about 10 meters from the Keraton Surya Negara or the palace indicating that the masjid and the palace cannot be separated (M. Kassim, 2014). Masjids built by the ruler is located facing the *alun-alun* or a aquare. The main architectural characteristic of ancient mosque of Indonesia is the tiered roof. Masjids built in various regions always includes elements of the region with its traditional values, illustrating the strong attachment to the traditional local culture. Islamic teachings were translated into the design of the masjid as illustrated by M. Kassim (2014):

"At the upper part of the tower and dome of the mosque there is a terraced roof shape and tapers to the top of which is a symbol of verticality and oriented to "the above", ie worship solely addressed to the Creator, the Highest, and were on top of everything. The building is considered the most sacred and most importantly have many levels and high roof. It gives the impression of the importance of the space beneath, a sacred space, commonly called cella." "While the pillar (column) are the pillars that surround the room cella (sacred space) and located in front of the mihrab space that contains a cosmological significance. Therefore, the imam as leader was to be followed by deeds. Meanwhile, all the same congregation or equal before Allah."

Masjid Agung Sumenep, Madura in Indonesia was built by the Sultanate of Sumenep. The builder who was responsible for the construction of the masjid was Lauw Pia Ngo, the grandchild of Lauw Khun Thing, one of the first six immigrants to arrive in the royal kingdom of Sumenep. His close relationship with the royal family (one of his family members was married to the royals of Sumenep) gave him the opportunity to work with the royal family during his early days. He was sent to Mataram by the royal kingdom to learn the architecture in the Mataram kingdom. Even with this new knowledge, his works are later defined to be inspired towards Chinese architectural influences as seen in the Masjid Agung Sumenep namely the construction, choice of colour and use of tiles from China. Masjid Agung Sumenep was built after the construction of Sumenep palace, one of the ancillary buildings of the Palace, which is as a place of worship for the royal family and society. The architecture is a mixture of Java, Madura and Chinese. The 13 pillars in the masjids is believed to represent the thirteen pillars of *solat* or Muslim prayer.

The Strait of Malacca is known for trading and countless numbers of merchants pass through this route, known as the "Spice Route". Some of them, coming from the Middle East and China, were Muslims. Mahfud (2014) writes that Cheng Ho, one of the key figures in China trading history, paved the way for introducing Chinese culture throughout its voyages. Coming through the trading route, some merchants would eventually settle down on the trading ports in order for them to trade their proceeds and stocking up their supplies and this would take a large number of days. Having to fare on sea, some for weeks and for some even for months, a large number of these sea merchants would settle down for good in the vicinity of the ports. They would set up a new community in the area thus exchanging cultural beliefs and values. During the declination of Majapahit, natives were attracted to Islam as the seafarers are on the idea of similarity, contrasting to the concept of Hindu social hierarchy (Sulasman, 2013). Van Leur (1960) pointed that this exchange paved the way for a "*politico*-economy" blueprint that caused many native merchants to believe in Islamic teachings and participated in trading with foreign Muslim merchant community. The social cultural exchanges also brought external influences on the architecture of the masjids as seen in the Masjid Agung Sumenep. Similarly, external influences on the design of Nusantara masjids is prevalent in Malaysia as seen in Masjid Kampung Hulu, Masjid Tengkera, and Masjid Kampung Keling which share similar characteristics. Melaka was an important port city in the Malay Archipelago that attracted many traders and merchants from all over the world.

Masjid Kampung Hulu, Melaka in Malaysia was built in 1728 by Dato' Samsuddin bin Arom, an immigrant from China, who himself was a Muslim convert and trader hired by the Dutch to supervise the construction of the masjid. The original building was later improved by Sheikh Al Omar bin Hussain Al-Attas. The masjid was inspired with Javanese architecture with a pyramidal three-tiered roof (Nor Afian, Hadzun and Narimah, 2014). The top layer was as a symbol of Iman towards God, the second layer was a symbol as relations between Man and the bottom layer is a symbol of relationship between nature and Man. Have three-tier roof, the top layer with the biggest area surface then the bottom layer. The masjid reflects the influence of Sumatera, Chinese, Hindu and Malay architecture. It has 4 main columns called *'tiang seri'* at the centre of the masjid. Originally the masjid was made of timber and was later renovated to brick walls in 1892. Most of the ornamentation were inspired by Chinese architecture including the usage of ceramic tiles brought from China and floral patterns widely found throughout the masjid, especially on the roof ridges. The masjid has the influences from Javanese, Chinese and Malay architecture. However, studies by the Malacca Museum Corporation disagree with the notion that the architecture of the old mosques in Malacca resembles a pagoda in China or bearing other foreign influences. A study carried out by local researchers found that carvings on the roofs and walls of the mosques that symbolize motifs of nature are indeed genuinely produced by Malay craftsmen (Nor Afian et. al., 2014).

Masjid Tengkera, Melaka was built during the early 17th century. However, there is no information on who was responsible for its construction. Based on its feature which resembles Masjid Kampung Hulu, it may have been by a Chinese builder. It features triple-tiered pyramidal roof and a square base, and is one of the few mosques in this country with a pagoda minaret. The minaret also functions as the lighthouse for the seafarers. The interior is a mixture of Malay, Chinese and Indonesian elements, with intricate carvings and wooden doors, windows, ceiling, beams, pulpit and four *tiang agung* (grand pillars). The masjid was inspired from Masjid Agung Demak in the Java, Indoensia. It has similar characteristics to Masjid Kampung Hulu; tiered roof, minaret and square plan for the main prayer area. It is based on the concept of four corners which relate to the four directions; north, South, West and East, similar to the idea of the Ka'aba which has four corners. The masjid main materials were made from timber imported from Kalimantan, Indonesia, which was later renovated into bricks in 1890. Its architecture is a mixture of Javanese, Chinese and Malay architecture.

The architecture of the masjid Nusantara in the Malay world were influenced by the stakeholders in different ways dependent on who was responsible for its construction; the one that decides to build, either the builder or the architect, or the one who oversee its construction. Masjids built by scholars normally started as a simple structure for preaching and eventually developed to cater for increasing number of people using the masjids for prayer and preaching. The masjids are simple and humble as compare to the one built by the rulers or individuals. The architecture of the masjid may reflect the local culture or have external influence when the scholar was from a different region such as in the case of some masjids in Malaysia built by scholars from Thailand. Masjids built by rulers normally were built within distance to the palaces which indicates the importance of the masjid in the life of the people. Its architecture was influenced by the master builder responsible for its construction, and because of this, the specifics of the design may be influenced by the origin of the master builder. The masjids built by merchant may have the most external influence and reflect a mixture of cultural influence of the local and the origin of the merchants. These findings in this paper is not conclusive given the limitation of data. Further studies are required on more cases before a conclusive finding.

4. Acknowledgement

This paper is made possible with the FRGS grant by the Government of Malaysia titled: Analysing Fundamental Values of Islam in the Planning and Design of Historic Nusantara Vernacular Masjid Architecture for Regional Sustainable Malay Islamic Architecture.

References

[1] Aoki, M. (1984). The Co-operative Game Theory of the Film. *Oxford: Clarendon Press.*

[2]*Business Dictionary*. (27 2, 2017). Retrieved from Business Dictionary: Stakeholder: http://www.businessdictionary.com/definition/stakeholder.html

[3]Carroll, A. B. (1991). The Pyramid of Corporate Social Responsibility: Towards the Moral Management of Organizational Stakeholders. *Business Horizons*, 39-48.

[4]Fontaine, C., Haarman, A., & Schmid, S. (December, 2006). The Stakeholde Theory.

[5]Friedman, A., & Miles, S. (13 9, 2006). Stakeholders: Theory and Practice. *Oxford University Press.*

[6]Harun, S. N. (2005). Perkembangan Sejarah dan Pengaruh ke atas Senibina di Malaysia: Sebuah Tinjauan Ringkas. *Buletin Perancang, JPBD, UiTM Perak.*

[7]*Investopedia*. (27 2, 2017). Retrieved from Investopedia: Stakeholder: http://www.investopedia.com/terms/s/stakeholder.asp

[8]Lee, K. M. (2008). It is Just Two Words! An Investigation of Two Cultures' Interpretations of Two New Curricula Terms. *International Journal of the Humanities . Jun2008, Vol. 6 Issue 3, ,* p117-121. 5p.

[9]Marsahid, C. M. (2014). Cheng Ho Mosque as New Silk Road Indonesia China Relations in Islamic Cultural Identity.

[10]Mastuhu. (1994). *Dinamika Sistem Pendidikan Pesantren.* Jakarta: INIS.

[11]Priyono, A. E. (1991). In Kuntowijoyo, *Paradigma Islam Interpretasi Untuk Aksi.* Bandung: Mizan.

[12]Sulasman, M. H. (2013). Rethinking the Historiography of Islam Indonesia. *Journal of Social Sciences (COES&RJ-JSS).*

Energy Efficiency Performance in Refurbishment Projects with Design Team Attributes As A Mediator: A Pilot Study

Siti Nor Azniza Ahmad Sekak[1], Prof. Dr Ismail Rahmat[2], Julitta Yunus[3], Sri Rahayu Mohd Saád[4], Mohd Hanafi Azman Ong[5]

[1]Centre of Studies for Quantity Surveying, Faculty of Architecture, Planning & Surveying Universiti Teknologi MARA, 40000 Shah Alam, Malaysia;
[2,3]Centre of Studies for Construction, Faculty of Architecture, Planning & Surveying Universiti Teknologi MARA, 40000 Shah Alam, Malaysia;
[4]Centre of Studies for Park & Amenity Management, Faculty of Architecture, Planning & Surveying
Universiti Teknologi MARA, 40000 Shah Alam, Malaysia;
[5]Faculty of Computers and Mathematical Sciences, Universiti Teknologi MARA Cawangan Johor
85009, Segamat,, Johor, Malaysia

syniza@salam.uitm.edu.my

Abstract. The Energy Efficiency (EE) plays an important role over the building life cycle and the implementation of EE in refurbishment projects has a significant potential towards the reduction of greenhouse gas emissions. However, the involvement of the design team at the early stage of the refurbishment projects will determine the success of EE implementations. Thus, a pilot study was conducted at the initial stage of the data collection process of this research to validate and verify the questionnaires.

Keywords: Energy Efficiency, Refurbishment projects, Design team attributes

1. Introduction

The Energy Efficiency (EE) improvement of building is one of the main decisions for saving primary energy and reducing greenhouse gas emissions [1]. The EE plays an important aspect over the building life cycle and energy efficient refurbishment in buildings has an enormous potential for energy saving towards the reduction of greenhouse gas emissions [2].
Elforgani & Rahmat [3] study revealed that the main environmental impact of a building is initiated during the early design stage. The early decisions made during design phase, will give impact on the overall project performance [4] and the effectiveness of a building to consume energy throughout its running life cycle is mainly determined at the early design stage [5].

2. Energy Efficiency Design Performance in Refurbishment Projects

In Malaysia, the government has been supporting the energy efficiency issues based on several initiatives including the regulatory matter to integrate the environmental issues into decisions during the design stage. These initiatives include The National Green Technology Policy (NGTP) in 2011, GBI (Existing Building Rating Tool), and Energy efficiency programmes such as Malaysian Industrial Energy Efficiency Improvement Project (MIEEIP), Small Renewable Energy Power Programme (SREPP and Building Energy Efficiency Programme (BEEP). All these efforts by the government showed the commitment towards energy efficiency development [6].
Nevertheless, the current initiatives functioning to encourage the energy efficiency building in Malaysia construction industry have been left behind schedule [7]. Reza et al. [8] stated that not many developers and designers in Malaysia are keen to move forward in the sustainable construction that include energy efficiency but instead remain with the conventional technique of construction even with the recent support from the government to build more sustainable buildings. Zainul Abidin [9] stated that there is a modest figure of sustainable projects in Malaysia and this showed that sustainability movement being slowly implemented among construction players and the energy efficiency movement is still at the early stage.

Content from this work may be used under the terms of the Creative Commons Attribution 3.0 licence. Any further distribution of this work must maintain attribution to the author(s) and the title of the work, journal citation and DOI.
Published under licence by IOP Publishing Ltd

3. Design Team Attributes

On the other hand, Horman et al., [10] pointed out that the integration working manner among the design team at the early stage of the project will determine the success in the energy efficient project performance. In refurbishment projects, according to Rahmat [11] the early design stage of the design team involvement will improve the project performance. Thus, more attention should be emphasized at this early point.

To embrace high performance in energy efficient design the design, teams have to work in an integrated manner with the right attributes. The good attribute of design team will enable them to interpret EE requirements [12]. Besides, the attributes of design team will influence the building refurbishment project performance [13]. Therefore, the attribute of the design team will determine the overall EE project performance and should be clearly identified.

4. Research Methodology

A pilot study was conducted for the data collection process. The questionnaires were distributed to the Architects, Electrical Engineers and Mechanical Engineers in the construction industry to validate and verify the questionnaire. By doing the pilot study the validity and consistency of the questionnaires can be confirmed. Wilson & McClean [14] pointed out that pilot study is important to look at the accuracy of the items in the questionnaire, as a guideline and to avoid confusion or difficulty in wording. Besides, the pilot study is important to receive a feedback on the validity of the items in the questionnaire, to get feedback on the presentation of the questionnaire and to get feedback on the format of the questionnaires which includes; layout, sectionalising, numbering or new items in the questionnaire.

The sampling of the questionnaires was distributed through the convenience sampling. This was done in the seminar on Energy Efficiency which was organized by Building Sector Energy Efficiency Project (BSEEP). Convenience sampling refers to the collection of information from members of the population who are conveniently available to provide it [15]. In this research convenience sampling may be the only choice available since the lack of respondents for refurbishment projects who involve in EE implementation. Besides, the convenience sampling is only meant for pilot survey of this research. In addition, the convenience sampling is only meant for pilot study of this research. During pilot study phase, the questionnaires were distributed to all participants in the seminar. 50 questionnaires were distributed to the participants. However, only 23 questionnaires were valid to be used for analysing puposes.

The questionnaire survey was divided into six sections which includes of general information, project information, design team attributes, and energy efficiency performance. Several different forms of scale were included in the questionnaires such as multiple-choice questions, five points Likert scale questions. To obtain higher response rate, the questionnaires are designed to be only four pages long. The questionnaires were developed based on the sources from many literature reviews involving various authors.

5. Data Analysis And Discussion

Based on Table 1, the majority of 82.6 percent of the respondents were engineers. These included of 39.1 percent were electrical engineers and 43.5 percent were Mechanical Engineer. The rest were architect (17.4%). Based on Table 1.2, almost half (47.8%) of the respondents had more than 10 years of experience in the construction industry. Those with 6 to 10 years' experience made up about one-third of the sample (34.8%). From these 23 respondents, about 60.9% involved in less than 5 refurbishment projects while only four of them involved in more than 15 refurbishment projects.

Table 1. Respondents' Job Title

Designation	**Percentage (N=23)**
Architect	17.4
Electrical Engineer	39.1
Mechanical Engineer	43.5
Total	**100.0**

Table 2. Number of refurbishment projects involved

Refurbishment projects	Percentage (N=23)
less than 5 projects	60.9
5 to 10 projects	21.7
11 to 15 projects	-
more than 15 projects	17.4
Total	**100.0**

Most (65.2%) of the refurbishment projects displayed in Table 3 had a contract value of less than RM 10 million. Only two refurbishment projects had a contract value more than RM 50 million while another six refurbishment projects companies managed to secure contract value between RM 11 million to RM 50 million. Thus, this pilot study has enabled to find of projects of varied size which would help produce a comprehensive result. The type of buildings involved for refurbishment project were mostly office (60.9%) and residential (30.4%). The percentage was calculated from multiple responses where in this question, the respondents can answer more than one. Apart from office and residential, industrial (13.0%), education (13.0%), hotel (8.7%) and shop (4.3%) were also involved (Table 4).

Table 3. Contract Value of Refurbishment Project

Contract value (RM)	Percentage (N=23)
less than RM2 million	30.4
RM2-10 million	34.8
RM11-50 million	26.0
more than RM50 million	8.7
Total	**100.0**

Table 4. Type of Building of Refurbishment projects

Type of building	Percentage (N=30)
Residential	30.4
Office	60.9
Industrial	13.0
Hotel	8.7
Shop	4.3
Education	13.0
Total	**100.0**

Table 5. Type of procurement system

Type of building	Percentage (N=23)
Traditional	43.5
Design and Build	52.2
No answer	4.3
Total	**100.0**

Table 6. Type of Client

Type of building	Percentage (N=23)
Private	65.2
Public	34.8

Total	**100.0**

Table 7. Green Building Certification

Type of building	Percentage (N=23)
Yes	34.8
No	65.2
Total	**100.0**

Table 8. Percentage of services work of the total contract value

Type of building	Percentage (N=23)
0%-10%	13.0
11%-20%	21.7
21%-30%	30.4
31%-40%	34.8
Total	**100.0**

Based on Table 5, the type of procurement system was almost equal between traditional (43.5%) and design and build (52.2%). Table 6 showed majority of the type of client was from the private sector (65.2%) compare to public sector (34.8%). Only 8 out of 23 refurbishment projects contained Green Building Certification. From these 8 projects, 5 had GBI, one had Green Mark and another one had LEED (Table 7). In terms of percentage of services work of the total contract value, 65.2% were between 21% to 40% (Table 1.8).

Table 9. Contract value and duration of the refurbishment project

	Min	Max	Median
Original contract value	RM0.5 million	RM100 million	RM6 million
Final contract value	RM0.5 million	RM100 million	RM8 million
Original contract duration	10 weeks	96 weeks	24 weeks
Duration for extension of time	0 weeks	144 weeks	15 weeks

Table 9 summarize the contact value and duration of the refurbishment projects. The original contract value was range from RM500, 000 to RM100 million with median value of RM6 million. This median value indicates that half of the companies had original contract value of less than RM6 million while the other half had value of more than RM6 million. The final contract value had the same range as the original except that the median value was slightly higher which is RM8 million. The duration for original contract value was range from 10 weeks to 96 weeks with the median of 24 weeks. There were companies who did not ask for time extension for the refurbishment projects. The maximum duration for extension of time was 144 weeks with half of the companies requested for less than 15 weeks of extension.

In measuring the consistency of a measure and to ensure that the items are assessing the same underlying construct, reliability analysis for Refurbishment Project Complexity, Design Team Attributes and Energy Efficient Performance were conducted. The data collected from the pilot study was verified through Cronbach's Alpha coefficient in order to ensure all items in the questionnaire are reliable and can be used for the final questionnaire survey. As a standard of reliability, the minimum acceptable standard value for Cronbach Alpha that is accepted is 0.7. Those above 0.8 represent good reliability and those above 0.90 represent excellent reliability value [16,17].

Table 4 indicates good estimates of internal consistency reliability as the values ranged between 0.92 and 0.95 for each construct. These values are far greater than the minimum level of 0.7, thus the questionnaire items indicate a satisfactory high degree of internal consistency and can be used for further statistical analysis.

Table 10. Reliability analysis

Attribute	Number of items	Cronbach Alpha
Refurbishment Project Complexity	19	0.92
Design Team Attributes		
Energy Efficient Performance	21	0.95
	18	0.93

6. Conclusion

In conclusion, the pilot questionnaire survey results provided important information about the EE design performance of refurbishment projects in Malaysia. The literature search and survey results showed that thorough studies into the refurbishment were relatively scarce. Thus, more detail study in refurbishment area is needed for Malaysian construction industry in the future.

Acknowledgements

The authors would like to thank LESTARI (600-IRMI/DANA 5/3/LESTARI (0156/2016) from RMC (Research Management Centre) of Institute of Research Management & Innovation (IRMI), Universiti Teknologi MARA, Malaysia for funding this research work.

Refferences

[1] S. Schimschar, K. Blok, T. Boermans, and A. Hermelink, "Germany's path towards nearly zero-energy buildings—Enabling the greenhouse gas mitigation potential in the building stock," *Energy Policy*, vol. 39, no. 6, pp. 3346–3360, Jun. 2011.

[2] E. Asadi, M. G. da Silva, C. H. Antunes, and L. Dias, "Multi-objective optimization for building retrofit strategies: A model and an application," *Energy Build.*, vol. 44, pp. 81–87, Jan. 2012.

[3] M. S. A. Elforgani and I. Bin Rahmat, "Green design performance of malaysian building projects - descriptive study," *ARPN J. Eng. Appl. Sci.*, vol. 6, no. 11, pp. 68–78, 2011.

[4] L. O. Oyedele and K. W. Tham, "Clients' assessment of architects' performance in building delivery process: Evidence from Nigeria," *Build. Environ.*, vol. 42, no. 5, pp. 2090–2099, May 2007.

[5] N. Ibrahim and A. Z. Ahmed, "Energy saving due to daylighting: a simplified rediction tool for wall envelope design of air-conditioned office buildings in Malaysia," *Built Environ. J.*, vol. 3, no. 1, pp. 63–75, 2006.

[6] C. C. Shing and H. O. Tick, "Review on Malaysia's national energy developments: Key policies, agencies, programmes and international involvements," *Renew. Sustain. Energy Rev.*, vol. 14, no. 9, pp. 2916–2925, Dec. 2010.

[7] M. S. A. Elforgani, "The influence of building project characteristics, quality of clients and design team attributes on green design performance measures adoption," Universiti Teknologi MARA, 2012.

[8] M. Reza, M. A. Marhani, R. Yaman, A. Arzlee, H. Noor, H. Noor, and H. Adnan, "Obstacles in Implementing Green Building Projects in Malaysia," *Aust. J. Basic Appl. Sci.*, vol. 5, no. 12, pp. 1806–1812, 2011.

[9] N. Zainul Abidin, "Investigating the awareness and application of sustainable construction concept by Malaysian developers," *Habitat Int.*, vol. 34, no. 4, pp. 421–426, Oct. 2010.

[10] M. J. Horman, D. R. Riley, A. R. Lapinski, S. Korkmaz, M. H. Pulaski, C. S. Magent, Y. Luo, N. Harding, and P. K. Dahl, "Delivering green buildings: process improvements for sustainable construction," *J. Green Build.*, vol. 1, no. 1, pp. 123–140, 2006.

[11] I. Rahmat, "The planning and control process of refurbishment projects," University College London, UK, 1997.

IOP Conf. Series: Materials Science and Engineering **291** (2017) 012022 doi:10.1088/1757-899X/291/1/012022

[12] Y. L. Yean, "Model for Predicting Performance of Architects and Engineers," *J. Constr. Eng. Manag.*, vol. 128, no. 5, pp. 446–455, 2002.

[13] A. S. Ali and P. A. Y. Cheong, "The designer in refurbishment projects: implications to the compatibility of design," *Struct. Surv.*, vol. 31, no. 3, pp. 202–213, 2013.

[14] N. Wilson and S. I. McClean, *Questionnaire design: a practical introduction*. University of Ulster, 1994.

[15] U. Sekaran and R. Bougie, *Research methods for business: A skill building approach*. United Kingdom: John Wiley & Sons, inc, 2009.

[16] J. F. Hair, G. T. M. Hult, C. M. Ringle, and M. Sarstedt, *Partial Least Squares Structural Equation Modeling (PLS-SEM)*. Sage Publications, Incorporated, 2014.

[17] J. C. Nunally and Bernstein, *Psychometric theory*. New York: McGraw-Hill, 1994.

IOP Conf. Series: Materials Science and Engineering **291** (2017) 012023 doi:10.1088/1757-899X/291/1/012023

An Evaluation on Factors Influencing Decision making for Malaysia Disaster Management: The Confirmatory Factor Analysis Approach

S N A Zubir[1], S Thiruchelvam[2], K N M Mustapha[2], Z Che Muda[1], A Ghazali[1] and H Hakimie[1]

[1]Institute of Energy Infrastructure (IEI), Universiti Tenaga Nasional,
Jalan IKRAM-UNITEN, 43000 Kajang, Selangor, Malaysia
[2]Chancellery, Universiti Tenaga Nasional,
Jalan IKRAM-UNITEN, 43000 Kajang, Selangor, Malaysia

E-mail: snaishahzubir@gmail.com

Abstract. For the past few years, natural disaster has been the subject of debate in disaster management especially in flood disaster. Each year, natural disaster results in significant loss of life, destruction of homes and public infrastructure, and economic hardship. Hence, an effective and efficient flood disaster management would assure non-futile efforts for life saving. The aim of this article is to examine the relationship between approach, decision maker, influence factor, result, and ethic to decision making for flood disaster management in Malaysia. The key elements of decision making in the disaster management were studied based on the literature. Questionnaire surveys were administered among lead agencies at East Coast of Malaysia in the state of Kelantan and Pahang. A total of 307 valid responses had been obtained for further analysis. Exploratory Factor Analysis (EFA) and Confirmatory Factor Analysis (CFA) were carried out to analyse the measurement model involved in the study. The CFA for second-order reflective and first-order reflective measurement model indicates that approach, decision maker, influence factor, result, and ethic have a significant and direct effect on decision making during disaster. The results from this study showed that decision- making during disaster is an important element for disaster management to necessitate a successful collaborative decision making. The measurement model is accepted to proceed with further analysis known as Structural Equation Modeling (SEM) and can be assessed for the future research.

1. Introduction

Natural disaster is commonly known to be disaster caused by the nature [1]. Natural disaster are beyond control of human beings and cannot be predicted accurately when it occur [2]. Major natural disasters consists of volcano eruptions, earthquakes, tsunami, avalanches, lahars (volcanic mudslides), landslides, blizzards, heat waves, hurricanes, typhoons, tornadoes, droughts, floods and others. No matter what types of natural disasters it may be, it usually leads to financial, environmental, and human loses [1], [2]. It is indeed a great danger for the earth if these sort of natural disasters tend to continues. Natural disasters are said to be cataclysmic or in other words a violent natural event that could give either a direct or indirect impact towards the public health and well-being [1].

Malaysia is safe from experiencing some of natural disasters that are happening around the world such

Content from this work may be used under the terms of the Creative Commons Attribution 3.0 licence. Any further distribution of this work must maintain attribution to the author(s) and the title of the work, journal citation and DOI.
Published under licence by IOP Publishing Ltd

as earthquakes, volcano, and typhoon because lies in a geologically stable region just outsides the "Pacific Ring of Fire" [1]–[5]. Hence, it is free from earthquakes and volcanic eruptions. It also lies too far south of the major typhoon paths, although tail-ends of tropical storms have occasionally hit it. However, that does not mean Malaysia is totally free from natural disasters and calamities, as it is often hit by floods, droughts, landslides, haze, tsunamis, and human made disasters [4]. The most intense natural disaster experienced in Malaysia is the flood [1], [2], [4], [5].

The Malaysia flood that happened from 15 December 2014 – 3 January 2015, has been described as the worst floods in decades. More than 200,000 people were affected and 21 people were killed due to this [1], [5]. It is important to reflect back what was happening in the flood 2014. It was unforgettable disaster happened at the end of 2014 [6]. By the end of 2014, most of the rivers in Kelantan, Pahang, and Terengganu had reached dangerous levels. Due to this, many businesses were affected and about 60,000 people were displaced by the following day. It is reported that Kelantan had the most evacuees with an estimation of 20,468 to 24,725 people followed by other states in Malaysia [1].

Disaster management is an alternative to improve resilience and therefore avoid or reduce the impact of natural disasters. Thus, disaster management is defined as an ongoing process composed of a set of activities before, during and after an event, separated into four main phases: mitigation, preparedness, response and recovery [7], [8]. In all these phases, information used by emergency agencies plays a key role in ensuring the reduction of impacts. Because of this, it is important that there are accurate, timely and complete information about the current state of environmental variables, as well as on the scientific predictions about the upcoming changes and their associated impacts [8].

The threat of natural and man-made disasters like earthquakes or deforestation is ever-present in our society. Past investigations have clearly shown that the number of natural disasters, as well as the victims and economic losses they cause is rising. This requires improvements in disaster preparedness, disaster mitigation, and organization of disaster responses. The emergency response process remains inefficient and sometimes weakly organized [7].

2. Background of the Study

From time to time, Malaysia experience disaster which affect the country and cause remarkable damages to lives and properties [3], [4], [9]. During the period of 1968-2004, Malaysia has experienced 39 disasters, classified as natural, human-made, and subsequent disasters that caused 1742 deaths and 2713 injuries [3], [4]. The matter raises the need for setting up for advanced warning systems, forecast future disasters, and effective disaster management in order to reduce the effect of such disasters [3], [4], [9].

Disaster management in Malaysia is not focused on a specific type of disaster. Every policy issued is applicable to all types of disasters, including flood [10]. After several disaster affecting Malaysia, the government has launched new mechanism for disasters and enhance the local knowledge, specialized rescue operation by establishing the Special Malaysian Disaster Assistance and Rescue Team (SMART), and National Security Council to be responsible about disaster management under the Directive No. 20 "The Policy and Mechanism of National Disaster Management and Relief" [3], [4]. National Security Council (MKN) Directive No. 20 was developed on May 11, 1997 [10]. It is merely a policy that is administered from the Prime Minister's Department to manage disaster generally [11]. The directive outlines the policy on disaster management and relief, including the roles and responsibilities of different agencies involved in managing disaster when occurred. The directive also set up emergency management programs to be well prepared and insure preparation, prevention, and mitigate the effect of disasters [3], [4]. In addition, the directive also covers the action of monitoring the activities of all agencies involved in disaster management. In 2012, the National Security Council have updated that Directive No. 20 for been revised to conform with the current changes, as well as the complexity of disasters is often the case at present [10].

A recurring problem in the management of response to natural disaster is the lack of coordination between various rescue agencies involved [12]–[14]. These agencies include not only the emergency services such as police, fire and rescue, and ambulance, but also local and national government bodies, private sector organizations and volunteer groups [14]–[16]. Successful operations are often attributed to effective inter-agency cooperation and coordination [14], [15].

Past investigations have clearly shown that the number of natural disasters, as well as the victims and economic losses they cause is rising. This requires improvements in disaster preparedness, disaster mitigation, and organization of disaster responses. The emergency response process remains inefficient and sometimes weakly organized. In order to handle disasters, save human lives, and reduce damages, it is essential to have quick response times, good collaboration and coordination among the parties involved, as well as advanced technique, resources, and infrastructure. The current response to disaster is inefficient and sometimes poorly organized. Various studies, research, and analyses have helped clarify needs, understand failures, and improve technology and organizational procedures. One of the major bottlenecks mentioned in many reports is communication between different parties involved in managing a disaster during the decision making process. The lack of good overview of the locations of teams, personnel, and facilities or insufficient information about the tasks needing to be completed may lead to misunderstanding and errors [7].

Decision making encompasses an array of processes some of which have been the subject of extensive investigations [17]. Decision making is a process of making a choice from a number of alternatives to achieve desired results. Anyone can make decisions. However, only a few people can make a good decision. In order to have a good decision, the decision makers should have a set of skills in decision making. During a disaster event, the main aim of the decision maker should be to save as many lives as possible. Therefore, decision making becomes an important element in everyday process and activities. Decision making carries a certain amount of responsibility and accountability. Decisions taken during disaster events impact a larger community and business audience. It also has impacts on community, economy, and the environment. Managing a response to catastrophic incident requires timely, effective decision making and a systematic management approach that applies sound tested principles [18].

During the decision making process, decision makers need to perceive the problem, gather relevant information, make the right judgment, and conduct a plan of action. Real-time situations such as crisis response or natural disaster relief, create a unique, threatening decision making environment that must be dealt with consistently in a timely manner [19]. Effective decision making under conditions of uncertainty involves the ability to recognize risk, formulate strategies for action, and coordinate with others in an effort to bring an incident under control quickly. Learning to make decisions effectively in urgent, uncertain conditions is not easily achieved [20].

3. Methodology

3.1 Sampling

The target respondents are among the lead agencies in two states in East Coast of Malaysia namely Kelantan Darul Naim and Pahang Darul Makmur. The agencies are National Security Council (MKN), Fire and Rescue Department (JBPM), Royal Malaysian Police (PDRM), Department of Irrigation or Drainage (JPS), Health Department (JK), Social Welfare Department (JKM), Public Works Department (JKR), and Malaysia Civil Defence Force (JPAM). The sampling strategy employed in this study are simple random sampling and cluster sampling. Simple random sampling was adopted so that everyone has an equal chance of being selected as part of the sample [21]. While cluster random sampling is a collective type of unit that includes multiple elements [22]. The study involved the selection of a suitable geographical area (East Coast of Malaysia) and the selection of a preferred agency in Malaysia disaster management. A total of 307 usable questionnaire had been obtained for further analysis.

3.2 Data Collection

Data collection was conducted over a two-month period during October and November 2015. The questionnaires are collected after one week it was distributed to each agency. It took ten working days at Kelantan and nine working days at Pahang. Before conducting surveys, each agency management's permission was obtained. A note-book was given to the participants as an incentive for participation in the study.

3.3 Data Analysis

Verifying the instrument for the study is done through Exploratory Factor Analysis (EFA) and Confirmatory Factor Analysis (CFA). At first, EFA was conducted first before proceeding to subsequent analysis which is CFA. EFA consists of principle component analysis (PCA) used for data reduction. The value at the threshold of 0.6 or above is suppresses once EFA procedure is applied. Once the authors execute the EFA procedure, this component will be employ in Structural Equation Modeling (SEM). [23] stated that SEM has two models namely measurement model (for CFA approach) and structural model (for path estimate). Every measurement of a latent construct needs to undergo CFA before modelling in SEM. The CFA was employed to validate the measurement model [24]. The assessment of fitness index comprised of three categories namely parsimonious fit, absolute fit and incremental fit but the researcher deserves to choose any fitness so as long the fitness chosen represents each category. However, the use of at least one fitness index from each category of model fit is suggested. The implementation of CFA ascertains the researchers to identify to what extent the strength of indicators and measurement model relates to decision making. Thus, the valid measurement model can help the researchers make a true interpretation based on their findings [23]. Table 1 below shows the fitness indexes with the threshold values.

Table 1. Fitness Indexes Categories of Model Fit and Level of Acceptance [25].

Name of Category	Name of Index	Level of Acceptance
Absolute Fit	Chi-Square	P-value > 0.05 Not applicable for sample size>200
	RMSEA	RMSEA < 0.08
	GFI	GFI > 0.90
Incremental Fit	AGFI	AGFI > 0.90
	CFI	CFI > 0.90
	TLI	TLI > 0.90
	NFI	NFI > 0.90
Parsimonious Fit	Chisq/df	Chisq/df < 3.0

4 Results

4.1 Descriptive Statistic

Prior to analysing the data, descriptive statistics were examined to check the normality of decision making.

Normally the data should be conducted to investigate how the standard of the data that has been collected so that the developing model may suit the parametric technique in the future research. Using Kolmogorov-Smirnov approach, the data is highly significant indicating as normal data. The reliability of the scale is determined by computing the coefficient alpha. The Cronbach's alpha of 0.6 or higher for a component reflects the measuring items under that particular component provide a reliable measure of internal consistency [26]. In this study, the Cronbach's Alpha values for decision making, approach, decision maker, influence factor, result, and ethic construct vary from 0.526 to 0.885. The result of reliability test shows that some of the sub-constructs from the main construct have to be deleted since the Cronbach's Alpha value less than 0.6.

4.2 Exploratory Factor Analysis

Using the data collected based on the sampling used, the data was executed using EFA to categorized which items is suited to be composed of the same component. The data was analysed using Statistical Package for Social Science (SPSS) version 22. The software used not only limited to the EFA procedure but has been used for the reliability to measure the goodness of actions. A PCA with varimax rotation was applied to analyse the data. [23] remarked that Kaiser-Meyer-Olkin (KMO) value greater than 0.60 were used to purify the measurement items. In this study, KMO for each variable (decision making, approach, decision maker, influence factor, result, and ethic) ranged between 0.747 to 0.861. This means the EFA approach is accepted and the result of KMO has achieved the requirement. This KMO value exceeds the recommended value of 0.6 and close to 1.0 with the Barlett's test significance value close to 0.0 suggest the data is appropriate to proceed with its reduction procedure [26]. The remaining items will proceed for the CFA approach.

4.3 Confirmatory Factor Analysis

CFA was analysed using Analysis of Moment Structure (AMOS) version 22. The CFA for second-order reflective and first-order reflective measurement model was performed with maximum likelihood estimate. The second-order model are Decision making (*DM*) and Influence Factor (*I*). While the first-order model are Approach (*Ap*), Decision Maker (*D*), Result (*R*), and Ethic (*E*). In the beginning, the items that carry the high factor loadings in EFA are deleted in measurement model for CFA approach. In order to fulfill the requirement of CFA specifically of the fitness index, reliability and discriminant validity, some of the items from each factor should be removed in the measurement model. Removal items are based on modification index (MI) that can be provided in Amos report to identify which items has carried the multicolinearity problem. Basically, MI over 15 should be considered as high multicollinearity problem.

The acceptable model is shown in Figure 1. The fit value for RMSEA = 0.060, CFI = 0.901, Chisq/df = 1.738, indicates that all indicators are significance. The model achieved the absolute fit, incremental fit, and parsimonious fit as shown in Table 2.

Table 2. The Assessment of Fit for Measurement Model.

Name of Category	Name of Index	Level of Acceptant	Comments
1. Absolute Fit	P-value	0.000	NA as the sample size > 200
	RMSEA	0.060	Meet the required level < 0.080
2. Incremental Fit	CFI	0.901	Meet the required level < 0.900
3. Parsimonious Fit	Chisq/df	1.738	Meet the required level < 3.0

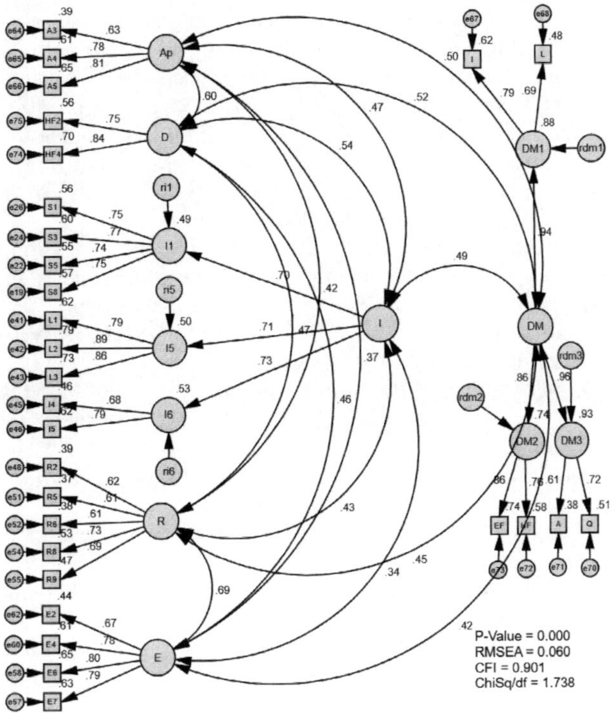

Figure 1. Measurement Model

5. Conclusion

Overall, the present study tries to explore the factor related to decision making in disaster management. In conclusion, based on the analysis, it shows that the data reasonably fit the model. This indicates that the model has achieved the absolute fit, incremental fit, and parsimonious fit that make up the achievement of the fitness indexes of the measurement model. The results revealed that approach, decision maker, influence factor, result, and ethic have a significant and direct effect on decision making during disaster. The measurement model is accepted to proceed with further analysis known as Structural Equation Modeling (SEM). The measurement model can be assessed for the future research so that next study would be included in this model.

Acknowledgement

The authors would like to take this opportunity to recognise the efforts and assistance from every individual and organisations on all aspects in the preparation of this paper. This study is supported by Ministry of Higher Education F2015103FRGS Grant. The authors also profoundly grateful to relevant agencies for their ideas, provision of relevant materials during the preparation and development whilst providing valuable input to complete this research.

References

[1] Islam R, Kamaruddin R, Ahmad S A, Jan S J, and Anuar A R, "A Review on Mechanism of Flood Disaster Management in Asia," *Int. Rev. Manag. Mark.*, vol. 6, no. 1, pp. 29–52, 2016.
[2] Wan Ahmad W I and Abdurahman S M, "Kelantan Flood 2014: Reflections from Relief Aid Mission to Kampung Kemubu, Kelantan," *Mediterr. J. Soc. Sci.*, pp. 340–344, May 2015.
[3] Hofstra H, Scholten H, Zlatanova S, and Scotta A, "Multi-user tangible interfaces for effective decision-making in disaster management," in *Remote Sensing and GIS Technologies for Monitoring and Prediction of Disasters*, S. Nayak and S. Zlatanova, Eds. Berlin, Heidelberg: Springer Berlin Heidelberg, 2008, pp. 243–266.
[4] Horita F E A and de Albuquerque J P, "An Approach to Support Decision-Making in Disaster Management based on Volunteer Geographic Information (VGI) and Spatial Decision Support Systems (SDSS)," *Proc. 10th Int. ISCRAM Conf. – Baden-Baden*, pp. 301–306, May 2013.
[5] Shafiai S and Khalid M S, "Flood Disaster Management in Malaysia: A Review of Issues of Flood Disaster Relief during and Post-Disaster," 2016, pp. 163–170.
[6] Elias Z, Hamin Z, and Othman M B, "Sustainable Management of Flood Risks in Malaysia: Some Lessons from the Legislation in England and Wales," *Procedia - Soc. Behav. Sci.*, vol. 105, pp. 491–497, Dec. 2013.
[7] Auf der Heide E, *Disaster response: principles of preparation and coordination*. St. Louis: Mosby, 1989.
[8] Denis H, "Coordination in a governmental disaster management meta-organization," *Int. J. Mass Emergencies Disasters*, vol. 131, pp. 25–43, 1995.
[9] Smith W and Dowell J, "A case study of co-ordinative decision-making in disaster management," *Ergonomics*, vol. 43, no. 8, pp. 1153–1166, Aug. 2000.
[10] Hidden A, *Investigation into the Clapham Junction railway accident*. London: H.M.S.O., 1989.
[11] Al-Omari A A, "The Relationship between Decision Making Styles and Leadership Styles among Public School Principals," *Int. Educ. Stud.*, vol. 6, no. 7, pp. 100–110, Jun. 2013.
[12] Withnaarachchi J and Setunge S, "Influence of Decision Making During Disaster and How It Impacts a Community," presented at the 10th International Conference of the International for Infrastructure Resilience and Reconstruction (I3R2), Purdue University, West Lafayette, Indiana, USA, 2014, pp. 176–188.
[13] Zhang D, Zhou L, and Nunamaker Jr J F, "A Knowledge Management Framework for the Support of Decision Making in Humanitarian Assistance/Disaster Relief," *Knowl. Inf. Syst.*, vol. 4, no. 3, pp. 370–385, Jul. 2002.
[14] Comfort L K and Wukich C, "Developing Decision-Making Skills for Uncertain Conditions: The Challenge of Educating Effective Emergency Managers," *J. Public Aff. Educ.*, vol. 19, no. 1, pp. 54–71, 2012.
[15] Chua Y P, *Mastering research methods*. Kuala Lumpur: McGraw-Hill Education, 2012.
[16] Christensen L B, Johnson B, and Turner L A, *Research methods, design, and analysis*. Pearson, 2015.
[17] Awang Z, Afthanorhan A, and Mohamad M, "An evaluation of measurement model for medical tourism research: the confirmatory factor analysis approach," *Int. J. Tour. Policy*, vol. 6, no. 1, pp. 29–45, 2015.
[18] Abdullahi M S, "The Effect of Finance, Infrastructure and Training on the Performance of Small and Medium Scale Enterprises (SMEs) in Nigeria: Exploratory Factor Analysis and Confirmatory Factor Analysis/Measurement Model," *Am. J. Innov. Res. Appl. Sci.*, pp. 67–86, Mar. 2016.
[19] Awang Z, *SEM Made Simple*, 1st ed. MPWS Rich Publication Sdn. Bhd., 2015.
[20] Awang Z, *Research Methodology and Data Analysis*. Uitm, 2012.

Comparative Study On The Ethical Perceptions Of Contractors And Designers In The China Construction Industry

Byung Gyoo Kang[1], Kaiwen Long[2], Cheng Zhang[3] and Jian Li Hao[4]
[1]University of Nottingham Ningbo China, Ningbo, China
[2, 3 and 4]Xian Jiaotong-Liverpool University, Suzhou , China

Byung-Gyoo.Kang@nottingham.edu.cn

Abstract. Ethics is becoming one of the most important requirements for successful business in the 21st century. The construction industry cannot be exceptional from this trend. However construction ethics management requires different approaches from other industries, as the products of the industry are construction projects which are completely different from mass production. Contractors and designers are two major participants in construction projects. The roles and responsibilities of these two project stakeholders decisively influence all aspects of construction project. Practically ethical perceptions of contractors and designers are one of main aspects to be considered for the effective and efficient management of ethics for the construction industry. This research has investigated the ethical perceptions of contractors and designers in the China construction industry. A questionnaire survey which contains 15 ethical issues and 6 demographic factors has been conducted. 170 construction professionals from construction companies and consulting companies have been participated in this survey. These 15 ethical issues are ranked in terms of seriousness, frequency and importance for both contractors and designers as a single group to understand the overall perceptions in the industry. The analysis has also been conducted for contractors and designers respectively to make comparisons between them. The top three serious ethical issues in the industry are 'Bribery and corruption', 'Failure to practice whistle-blowing' and 'Improper bidding practices'. Contractors and designers showed similar outcomes for the top five important ethical issues. But for the middle ranged ethical issues, they showed some differences. Further researches are required to identify the causes of the similarity and differences.

Keywords: Construction, Ethics, Contractor, Designer

1. INTRODUCTION

Back to the Palaeolithic Age between 40,000 and 12,000 B.C., when caves and simple structures were used for human beings to live, construction started to play an important role in human history [1]. Since then, construction has been one of the most labour intensive industries [2]. The levels of ethical perceptions of construction professionals should be one of top priorities of ethics management in the construction industry. The consequences of unethical behaviours of contractors and designers can causes serious issues for successful completion of construction projects. The primary stakeholders of construction projects are clients, designers, contractors, sub-contractors, material suppliers and end-users [3]. Particularly, designers and contractors are the two main parties who eventually produce the quantity and quality of the contract. However in the traditional design-bid-build type of contracts, these two parties are often involved in dispute mainly because the designers work as the representative of the clients. Even though new procurement systems such as Integrated Project Delivery (IPD), PPP/PFI (Public Private Partnership/ Private finance Initiative) and more often design-build have been introduced to the construction industry, still design-bid-build is the most common type of project delivery method, especially in China. Therefore it is worth investigating the ethical perception of

Content from this work may be used under the terms of the Creative Commons Attribution 3.0 licence. Any further distribution of this work must maintain attribution to the author(s) and the title of the work, journal citation and DOI.
Published under licence by IOP Publishing Ltd

designers and contractors in the China construction industry. The objectives of this research are to investigate the ethical perceptions of contractors and designers in the China construction industry.
To compare the ethical perceptions between contractors and designers with respect to the frequency and seriousness of ethical issues.
The China construction industry has been chosen to be investigated in this research, considering the size and role of the industry in the international construction markets. Further, Chinese contractors have become more active when participating in international construction projects [2]. So the outcomes of this research will provide a good foundation for the future research for the ethics management in the international construction projects.

2. Literature Review

Ethics

Ethics which is a branch of philosophy can be defined as 'philosophical inquiry into the nature and grounds of morality [4]. It is moral philosophy or philosophical thinking about morality, moral problems and moral judgements. For construction ethics management, basically three major theories of ethics need to be studied - virtue ethics, deontology and utilitarianism. These ethical theories can be used to develop code of ethics, ethics training programme and ethical decision making process [5].

Virtue Ethics

Virtue is defined as a character trait that contributes to one's flourishing, such as generosity, integrity, honesty and patience [6]. Virtue ethics is the ethical theory and tradition that focuses on finding which character traits are the most important for living an ethically good life [7]. The major concepts for virtue ethics are to develop good character, such as fairness, courage, and self-control, and it focuses on how such excellences help us live good lives, treat ourselves and others well, and share thriving communities. The leading philosophers are Socrates, Plato and Aristotle.

Deontology

The theory and tradition of deontology focuses on ethical duties and whether actions are ethically right or wrong depends on fulfilment of one's ethical duties. It focuses on the means rather than the ends. Therefore, deontological ethics can be classified as a non-consequentialist ethic [8]. The deontologists argue that there are some things that should not be done even though they maximize the total utility. The leading philosopher is Emmanuel Kant.

Consequentialism/Utilitarianism

According to the theory and tradition of consequentialism, actions (including rules, laws or standards) ought to be judged to be ethically right or wrong depending on the consequences. Therefore in consequentialism, the moral value of and action simply depends on the results. Utilitarianism is the most dominant theory for ethical decision making in engineering fields as it provides clear and objective directions. A problem with utilitarianism is that it just considers the total utility that is produced in the society and fails to take into account how to distribute the utility among the society's members. One of the common processes to deduce conclusions in utilitarianism is cost-benefit analysis which civil engineers and constriction engineers are familiar with [5]. The leading philosophers are Jeremy Bentham and John Stuart Mill.

Construction Ethics

The construction industry plays a substantial role in a country's national economy, irrespective of the country's levels of economy development. The construction industry requires a different approach to ethics management from manufacturing or factory based industries. In manufacturing or factory based industries, mass production methods are common practice and stakeholders are related to business However, in the construction industry, most projects tend to be one-off and the stakeholders are related to project. The major participants in construction industry are the clients, designers, contractors and supplier [3]. Table2 show 15 common ethical issues in construction industry [9]. These ethical issues were identified by Jackon (2004) and used successfully for construction management researches in USA, UK, Korea and Malaysia [3], [5], [9]. In this research, these 15 ethical issues have been used to understand the perceptions of designers and contractors in the China construction industry.

Table 1. most common ethical issues in the construction industry

	Issues	Examples
1	Abuse of client resources	Over billing for time and material, inflating hours, excessive changes and charges, etc
2	Abuse of company resources	Abuse of travel allowance, fudging on time cards, personal use of company supplies, equipment, telephone, or facilities, using company employees for personal projects or benefit, etc.
3	Alcohol and drug abuses	Use of alcohol or drugs while on the job, excessive use of alcohol or drugs while off the job, effects of substance abuse on performance and decision-making.
4	Improper political or society involvement, conflict of interest	Political contributions or activity for personal or company gain, undue influence, fraud, conflicts of commitment, financial, personal, political, or other interest in people or organizations that one performs construction services for, etc.
5	Favouritism, discrimination and harassment	Unfair treatment on the basis of race, sex, etc, in business, or in relative to evaluations, promotions, or recommendations, supervisory harassment of subordinates, sexual harassment, etc.
6	Lack of protection to public's health	Poor safety or risk analysis or assessment, neglect in regard to worker safety, hazardous materials, natural hazards, etc.
7	Lack of protection to the environment	Conduct contributing to pollution, deterioration or destruction of air, water, or nature, resource depletion, etc.
8	Failure to practice whistle-blowing	Falsely blaming others for poor performance or schedule delays, company disloyalty, company communication, reporting, and grievance procedures, improper punishment or retaliation against an employee, etc.
9	Improper bidding practices	Collusion, Bid-shopping, bid peddling, etc.
10	Bribery and corruption	Excessive gifts, entertainment, or gratuities, undue influence, inside information, failure to maintain independent judgment; kickbacks, bribery or blackmail, fraud, etc.
11	Mishandle sensitive information	Revealing or obtaining proprietary or confidential information, revealing or discussing confidential bids and prices, misrepresentation of data, violation of privacy, gossip, etc.
12	Misrepresentation of completed work or value of work	Inflating completed work percentages, front-end loading schedules of value, etc.
13	Misrepresentation of financial status or records	Misinforming or misleading the IRS, lending institutions, banks, clients, etc.
14	Lack of quality or quality control of work	Cutting corners in the face of budget or time pressures, not satisfying specifications, hedging on standards, not performing in a workmanlike manner, etc.
15	Lack of competence or misinterpretation of competence	Operating outside one's area of experience or expertise, operating without a license, misleading advertisement for performance or products, misleading information on resumes or pre-qualification statements, etc.

3. Research Methodology

This survey participants are the professionals who work in the China construction industry. The participants work in construction companies or consulting companies in the southwest and the east of China. During the survey, anonymity was guaranteed to minimise the passive replies from the participants as ethics is a sensitive area in the industry. The method of sending questionnaire was using email and social software to communicate with the participants. The questionnaires were sent to the head of companies and distributed to the employees. The progress of survey was confirmed by phone calls. All the questionnaires were collected immediately when it was finished. The process of questionnaire survey took about one month. The questionnaire is composed of three parts. The first part is the demographic information about participants such as gender, age, job, working experience, education level, the existence of company codes of ethics and the effectiveness of the codes of ethics.

The second part is about the seriousness of 15 ethical issues and the third part of questionnaire is about the frequency of 15 ethical issues with responses using Likert Scale of 1 to 5.

4. Findings And Discussion

The number of contractors participated in this questionnaire survey was 94 and the number of designers was 76. The designers include consulting engineers. Although the numbers of these two professions are not equal, considering the number of participants, the results of the analysis can be reliable to understand the ethical perceptions between designers and contractors.

Mean value, standard deviation and rank of these 15 ethical issues based on seriousness and frequency from contractors and designers are presented below. The rank of these ethical issues is based on the mean value of each ethical issue. In the rank term, 1 represents the highest mean value and 15 represent the lowest mean value. If two issues have same mean value, the standard deviation would be used to decide the ranking. The lower value of standard deviation represents that the data are close to the mean value. The higher value of standard deviation represents that the data have a large difference from the mean value. Therefore, if two issues have same mean value, the issue has a lower value of standard deviation can be ranked higher. In this research, the seriousness of 15 ethical issues with responses uses Likert Scale of 1 to 5, where 1 indicates ‘extremely serious’, 2 indicates ‘very serious’, 3 indicates ‘serious’, 4 indicates ‘little serious’ and 5 indicates ‘not serious at all’. The frequency of 15 ethical issues with responses uses Likert Scale of 1 to 5, where 1 indicates ‘very often’, 2 indicates ‘often’, 3 indicates ‘sometimes’, 4 indicates ‘rarely’ and 5 indicates ‘never’. The values of importance are calculated by multiplying the value of mean of seriousness and the value of mean of frequency in each ethical issue.

Table 2 shows the overall ethical perceptions combining both contractors and designers. The issue of bribery and corruption ranks at the 1st in seriousness part. It means that the issue of bribery and corruption is the most serious ethical issue in China construction industry. The issue of lack of quality or quality control of work ranks at the 1st in frequency part. It means that the issue of lack of quality or quality control of work is the most frequent ethical issue from contractors and designers in China construction industry. According to the importance part, the most important ethical issue is the bribery and corruption.

Table 2. Seriousness, frequency and importance of ethical issues (both contractors and designers)

Ethical issues	Seriousness			Frequency			Importance	
	Mean	Std. Deviation	Rank	Mean	Std. Deviation	Rank	Value	Rank
Bribery and corruption	2.67	1.282	1	3.32	1.175	5	8.864	1
Lack of quality or quality control of work	3.34	1.172	5	2.72	1.273	1	9.085	2
Abuse of client resources	3.42	1.17	7	3.27	1.239	4	11.183	3
Improper political or society involvement, conflict of interest	3.55	1.177	12	3.19	1.198	2	11.325	4
Improper bidding practices	3.23	1.376	3	3.52	1.056	10	11.37	5
Failure to practice whistle-blowing	3.19	1.226	2	3.61	1.062	13	11.516	6
Abuse of company resources	3.59	1.075	13	3.25	1.367	3	11.668	7
Mishandle sensitive information	3.43	1.145	8	3.44	1.125	8	11.799	8
Favouritism,	3.54	1.11	11	3.42	1.118	6	12.107	9

discrimination and harassment								
Misrepresentation of completed work or value of work	3.44	1.151	9	3.57	1.145	12	12.281	10
Lack of protection to public health, safety and welfare	3.52	1.056	10	3.53	1.11	11	12.426	11
Misrepresentation of financial status or records	3.41	1.154	6	3.65	0.999	14	12.447	12
Lack of competence or misinterpretation of competence	3.65	0.999	14	3.42	1.17	7	12.483	13
Lack of protection to the environment	3.26	1.233	4	4.14	0.991	15	13.496	14
Alcohol and drug abuses	4.15	0.985	15	3.45	1.12	9	14.318	15
Average of Mean	3.426	0.864		3.434	0.85			

Table 3 compares the ethical perception between contractors and designers. The issue of 'bribery and corruption' ranks as the most serious and the most important issue for both contractors and designers. The most frequent ethical issue from contractors and designers is 'lack of quality or quality control of work'. Contractors and designers showed the same rankings of importance in 7 issues out of 15 (highlighted). The first, second, fourth and fifth important issues are the same between contractors and designers. Therefore it can be concluded that contractors and designers have similar perceptions for the major ethical issues. However they showed some differences in the middle ranked ethical issues.

Table 3. Seriousness, frequency and importance of ethical issues (contractors vs. designers)

	Contractors						Designers					
	Seriousness		Frequency		Importance		Seriousness		Frequency		Importance	
	Mean	Rank	Mean	Rank	Value	Rank	Mean	Rank	Mean	Rank	Value	Rank
Bribery and corruption	2.6	1	3.55	7	9.23	1	2.76	1	3.04	3	8.39	1
Lack of quality or quality control of work	3.6	9	2.61	1	9.396	2	3.03	3	2.87	1	8.696	2
Improper bidding practices	3.3	3	3.57	10	11.781	3	3.14	6	3.46	13	10.86	8
Improper political or society involvement, conflict of interest	3.71	12	3.24	2	12.02	4	3.36	11	3.13	5	10.517	4
Failure to practice whistle-blowing	3.24	2	3.79	13	12.28	5	3.12	4	3.38	10	10.546	5
Lack of protection to public health, safety and welfare	3.57	8	3.45	4	12.317	6	3.45	13	3.63	14	12.524	13

Favouritism, discrimination and harassment	3.47	5	3.57	9	12.388	7	3.63	14	3.22	7	11.689	12
Mishandle sensitive information	3.56	7	3.49	5	12.424	8	3.26	8	3.37	9	10.986	10
Abuse of company resources	3.79	13	3.3	3	12.507	9	3.34	9	3.18	6	10.621	6
Abuse of client resources	3.67	11	3.49	6	12.808	10	3.12	5	3	2	9.36	3
Misrepresentation of financial status or records	3.46	4	3.82	14	13.217	11	3.34	10	3.45	12	11.523	11
Misrepresentation of completed work or value of work	3.62	10	3.72	12	13.466	12	3.22	7	3.38	11	10.884	9
Lack of competence or misinterpretation of competence	3.82	14	3.67	11	14.019	13	3.43	12	3.12	4	10.702	7
Lack of protection to the environment	3.49	6	4.06	15	14.169	14	2.99	2	4.22	15	12.618	14
Alcohol and drug abuses	4.09	15	3.57	8	14.601	15	4.24	15	3.3	8	13.992	15
Average of Mean	3.53		3.53									

5. CONCLUSION

This research has investigated the ethical perceptions of contractors and designers in the China construction industry. With respect to seriousness, frequency and importance, 15 ethical issues were analysed. The top three serious ethical issues in the industry are 'Bribery and corruption', 'Failure to practice whistle-blowing' and 'Improper bidding practices'. The top three frequent ethical issues are 'Lack of quality or quality control of work', 'Improper political or society involvement', 'conflict of interest' and 'Abuse of company resources'. The top three most important ethical issues are 'Bribery and corruption', 'Lack of quality or quality control of work', 'Abuse of client resources'. Both seriousness and frequency are reflected in calculating the importance. In the comparisons between the ethical perceptions, contractors and designers showed similar outcomes for the top five important ethical issues. But for the middle ranked ethical issues, they showed some differences. Further researches and analysis will be required to identify the causes of these similarity and difference between contractor and designers.

References

[1] Zhou, Z. P., Goh, Y. M., Li, Q. M. 2015. Overview and analysis of safety management studies in the construction industry. Safety Science, 72: 337-350.

[2] Du, L, Tang, W. Z., Liu, C. N., Wang, S. L., Wang, T. F., Shen, W. X., Huang, M. and Zhou, Y. Z. 2016. Enhancing engineer-procure-construct project performance by partnering in international markets: Perspective from Chinese construction companies. International Journal of Project Management, 34: 30-43.

[3] Kang, B. G. and Shahary, M. A. B. 2012. Ethical perception and demographic factors: A comparative study between contractors and designers in the Malaysia Construction Industry. Malaysian Construction Research Journal, 10(1).

[4] Feldman, F. 1978. Introductory ethics. Prentice-Hall.

[5] Kang B.G. 2009. Principles and Practices of Construction Ethics Management: With a Comparative Study between the UK and Korea, VDM

[6] McBrayer, J. P. and Markie, P. J. 2014. Introducing ethics: a critical thinking approach with readings. Oxford University Press.

[7] Hutchings, K. 2010. Global ethics: an introduction. Polity.

[8] Mizzoni, J. 2010. Ethics: the basis. Wiley-Blackwell.

[9] Jackson, B 2004. The perceptions of experienced construction practitioners regarding ethical transgressions in the construction industry, The International Journal of Construction Education and Research, 1(2).

Sedimentary architecture and depositional environment of Kudat Formation, Sabah, Malaysia

Samira Ghaheri [1], Mohd Suhaili [2*], Nasiman Sapari[3], Mohammadsadegh Momeni[4]
[1&2] Department of Geosciences, University Teknologi PETRONAS, 32610 Seri Iskandar, Perak Darul Ridzuan, Malaysia
[3]Department of Civil and Environmental Engineering, University Teknologi PETRONAS, 32610 Seri Iskandar, Perak Darul Ridzuan, Malaysia
[4]Department of Petroleum Engineering, University Teknologi PETRONAS, 32610 Seri Iskandar, Perak Darul Ridzuan, Malaysia

suhali.ismail@utp.edu.my

Abstract. Kudat Formation originated from deep marine environment. Three lithofacies association of deep marine turbidity channel was discovered in three Members of the Kudat Formation in Kudat Peninsula, Sabah, Malaysia. Turbidite and deep marine architecture elements was described based on detailed sedimentological studies. Four architecture elements were identified based on each facies association and their lithology properties and character: inner external levee that was formed by turbidity flows spill out from their confinement of channel belt; Lobes sheet that was formed during downslope debris flows associated with levee; Channel fill which sediments deposited from high to low density currents with different value of sediment concentration; and overbank terrace which was formed by rapid suspension sedimentation. The depositional environment of Kudat Formation is shelf to deep marine fan.

Keywords: turbidite system, deep marine, facies association, architectural elements, Kudat Formation, Sabah

1. Introduction

Turbidite sediments in deep marine environment are common in most part of the world (e.g. Gulf of Mexico, Mississippi Fan, etc.). Turbidite deposits are form by density currents composed of water and suspended sediments. Documentation and study of sedimentation processes and deposition that occur in deep water turbidites are important due to prolific hosts for oil and gas resource [1]. In deep marine deposition study geoscientist often use seismic data, but due to deep water complexity of deposition seismic data alone are mostly insufficient. The outcrops study is required to identify the architectural elements, which control the reservoir or source rock physical and geochemical properties. During the last decade many authors had worked on detailed sedimentological architectural elements, e.g., [2-16]. In terms of hydrocarbon exploration and field development study, Kudat Peninsula is located Northwest Sabah (Figure 1) near to offshore oil and gas fields and became noteworthy. This area also had some natural oil seepages occurrence [17].

Content from this work may be used under the terms of the Creative Commons Attribution 3.0 licence. Any further distribution of this work must maintain attribution to the author(s) and the title of the work, journal citation and DOI.
Published under licence by IOP Publishing Ltd

IOP Conf. Series: Materials Science and Engineering **291** (2017) 012025 doi:10.1088/1757-899X/291/1/012025

This paper attempts to explain the sedimentological characteristics and facies architecture which represent the key elements for the identification of depositional environment of Kudat Formation which is located near to the oil and gas production field.

2. Geological Setting

Sabah region is tectonically active and complex. The region is surrounded by three seas (South China Sea, Celebs Sea, Sulu Sea) and is affected by tectonic activity of these seas [18].
The study area is located in the Northwest part of Sabah (Figure 2). Field work was carried out to study the rock formation of the Sikuati Member in which the oil seepage was found in Kampung Minyak.
Kudat Formation and Crocker Formation covered almost entirely the North part of Sabah. The Kudat Formation age between Lower to Middle Miocene and It has three members: Tajau, Sikuati and Gomantong [19]. Tongkul [20, 21] interpreted during Middle Eocene to Middle Oligocene a basin was formed in the area due to south-eastward subduction of the South China Sea under the western part of the Celebes Sea. The Kudat Formation was deposited in central part of this basin as deep to shallow marine deposition.

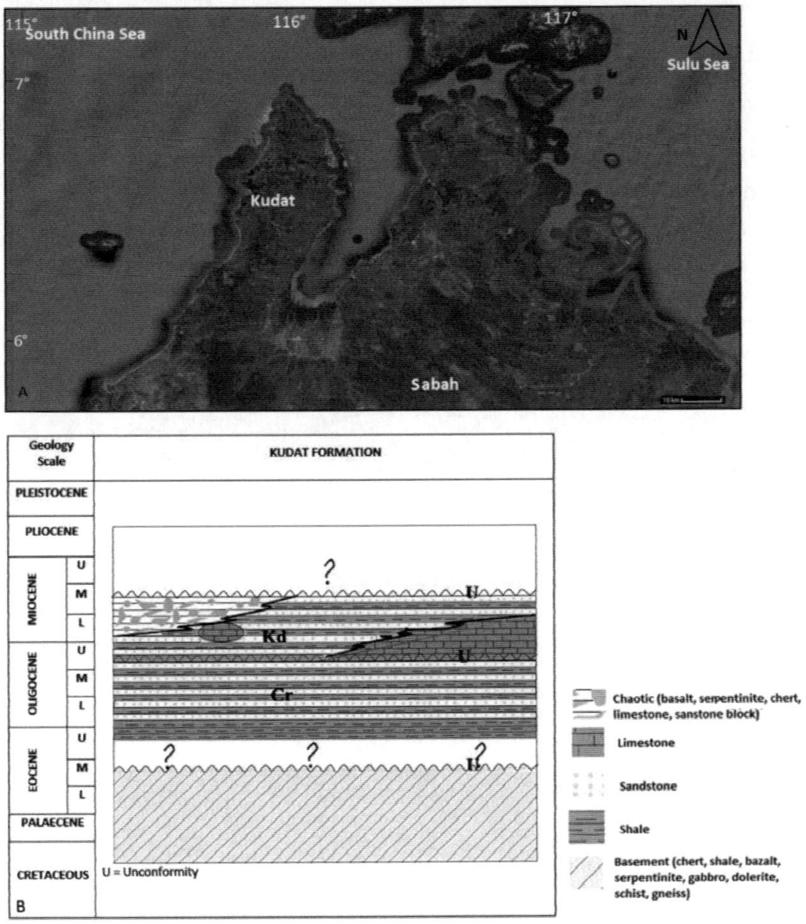

Figure 1: A) The location map of study area (Kudat Peninsula). B) Stratigraphic framework of Kudat Formation [18, 22, 23].

The Kudat Formation can be divided into three lithological units: the first one is lower sandy unit that consist of sandstone and mudstone with the sandstone predominating. Middle Muddy Unit comprises sandstone and mudstone with the mudstone dominance and the third one is upper sandy unit include sandstone and mudstone [24]. The lower sandy unit was deposited in the northern part of the Kudat Peninsula and coincides with the Tajau Member. The middle muddy unit which is known as the Sikuati Member covered the middle part of Kudat Peninsula where Kampung Minyak is located. The last unit occurred in the southern part of Kudat Peninsula and coincides with Gomantong Member [19].

3. Methods

Forty-eight outcrops were investigated from the North to South part of Kudat Peninsula and stratigraphic sections of Kuadt Formation were logged and described (Figure 2). Sedimentary units are defined collectively by physical and biogenic properties, lithology, texture and sedimentary structures. Logging of the lithology and sedimentology properties were used to provide vertical and lateral variation in texture (grain size, sorting and roundedness), lithological, physical and biogenic sedimentary structures, fossil content, bed thickness, and nature of bedding contact. The vertical logging was used as the basis for the interpretation of facies distributions, geometries and depositional environments at scales from the individual architectural element to the complete slope–turbidite and debris systems. The lithofacies transition patterns were used to interpret the paleoenvironment of deposition following the method of Pemberton and Mayll [4, 25]. In addition, there is a difficulty to find proper outcrop in terms of size due to high vegetation density in Malaysia.

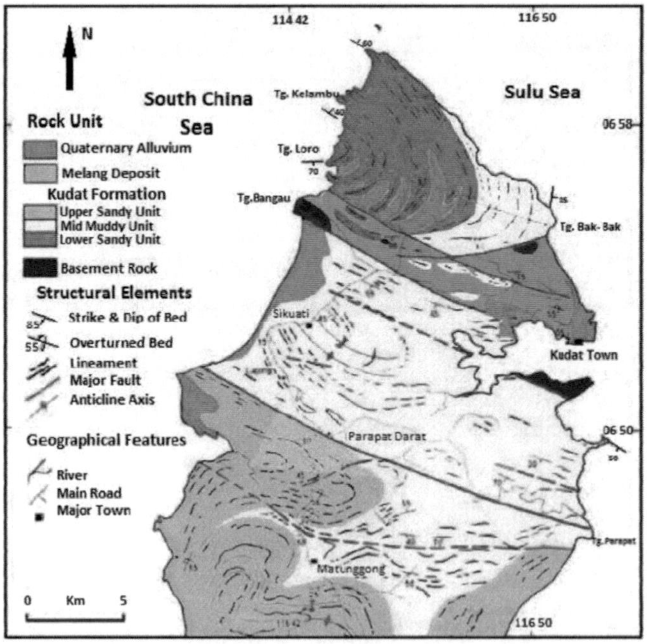

Figure 2: General Geology of the Kudat Peninsula (Modified from Tongkul, 2006)

4. Facies Association and Depositional Environment Interpretations

The Kudat formation in total was subdivided into three members. The Lower Sandy Member including one facies association, the Middle Muddy Member consists three facies association and the last member, The Upper Sandy is subdivided into two facies association.

4.1. *The Lower Sandy Member (Tajau)*

This member located at northern part of Kudat Peninsula and its strata is about ~ 409 m along ~ 54.2 Km2. In the field work investigation strata of the Lower Sandy Member Nine Outcrops (Figure 3) were

IOP Conf. Series: Materials Science and Engineering **291** (2017) 012025 doi:10.1088/1757-899X/291/1/012025

measured in detail and one associated facies (LSFA1) were identified (Figure 4). Field based lithological description analysis was used to characterize each facies (Table 1).

4.1.1. LSFA1: Thick and medium bedded sandstone and mudstone (LSFA1)

Description: This facies association includes five facies and comprises interbedded different sandstone facies (A, B, C and E) and mudstone (D) (Figure 4). Bed sets of facies association range from ~10 cm to ~ 5 m thick. Bed bases are commonly flat to undulating contacts with shallow scours up.

Facies A generally is observed at the base of strata and includes three sections and comprises thin and medium bedded pebbly sandstone. The sections consist of aggregation of pebbles (AP), two or three sets of sandstone and siltstone laminations (SL) and non-aggregation pebbly sandstone (NAP) (Figure 5A). The pebbles roundness ranges from angular to sub-angular and their size ranges between 2 to 6 mm. Bed sets of facies range from <80 cm to ~ 190 cm thick. Bed bases are flat to erosional with very shallow scours. Sandstones beds are generally light grey in colour.

In some beds the structures are commonly deformed and showed dewatering structures such as dish structures in non-aggregation pebbly sandstone (Figure 5B). In some outcrops at the base of pebbly sandstone, even rare the coal pebble and mud drops are present. In one outcrop hummocky cross like in different scale occurred in pebbly sandstone (Figure 5C). Other common sedimentary structures in this facies are convolute structure (Figure 5D) and loading. Facies B and C commonly over lied on pebbly sandstone (facies A). Facies B (iron trace sandstone) consists non-continue iron trace lamination of coarse to medium size sand. The thickness of beds ranges between 80 to 200 cm. Beds based are commonly erosional with shallow scours. Sandstone are generally light grey, and the sands are rounded and well sorted. Weathering structures such as honeycomb are abundant in this facies (Figure 5E), especially near to the coast.

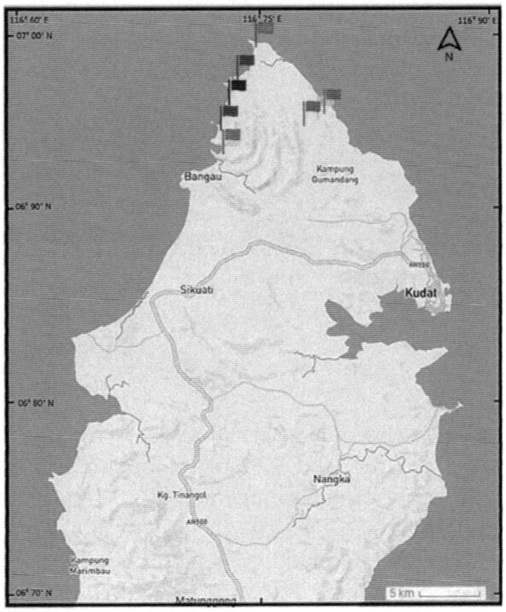

Figure 3: Outcrops location of the Lower Sandy Member

Lamination sandstone (facies C) comprises very coarse to coarse sandstone units of undulated to straight laminations. This facies syndepositionally deformed and consist dewatering structures. The thickness of planner lamination varies between 1 to ~ 7 cm.

IOP Conf. Series: Materials Science and Engineering **291** (2017) 012025 doi:10.1088/1757-899X/291/1/012025

Table 1: Facies description of Lower Sandy Member

Name	Description	Facies
Pebbly sand (Facies A)	Including three sections and comprises thin and medium bedded pebbly sandstone. The sections consist of aggregation of pebbles (AP), two or three sets of sandstone and siltstone laminations (SL) and non-aggregation pebbly sandstone (NAP)	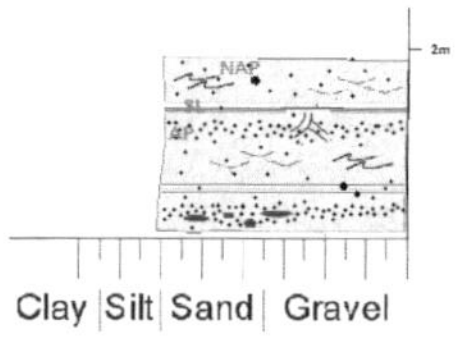
Iron trace sandstone (Facies B)	Consisting non-continue iron trace lamination of coarse to medium size sand.	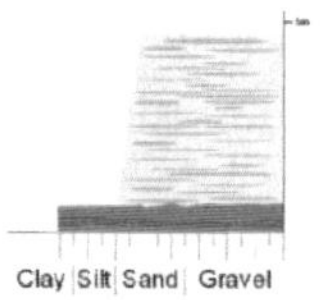
Lamination sandstone (Facies C)	Comprising very coarse to coarse sandstone units of undulated to straight laminations. This facies syn-depositionally deformed and consists dewatering structures.	
Mudstone (Facies D)	Mudstone beds including carbonaceous laminations and thickness of beds range from 10 to 28 cm. This facies.	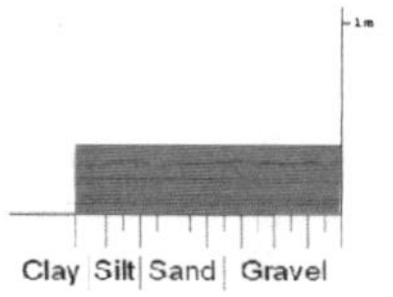
Structureless sandstone (Facies E)	Comprising very coarse to coarse graded structureless sandstone. The sands grain are well rounded and well sorted.	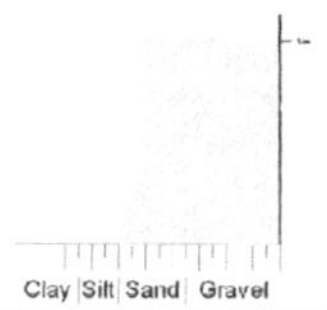

The sand grains are sub rounded to rounded and moderate sorted. The beds based are commonly undulating with shallow scours up and the sandstone beds have a distinctly light grey. In a small number of beds, coal pebbles although uncommon, occurred. In some location a single or few sets of thin overlying small-scale cross sets (Figure 5F) (as single 3-6 cm thick) occurred at the top of lamination sandstone. The present of flame (Figure 5G) and load structures are most common in strata of this facies. In one outcrop the hummocky cross like and convolute structure are presented. In one outcrop the hummocky cross like and convolute structure are present in this facies. In some outcrops some stone balls (Figure 5H) or their trace are presented and showed that they influenced the flow directions. These stone balls are generally present in sandstones facies and their size range from 5 cm to 70 cm.

Generally, the mudstone bedding (facies D) is overlain abruptly the facies B, C and E. Its thickness ranges from 10 to 28 cm. This facies includes carbonaceous laminations. The planner stratified structures are commonly syndepositionally deformed and faulted. At the top of strata the structureless sandstone (facies E) was deposited.

IOP Conf. Series: Materials Science and Engineering **291** (2017) 012025 doi:10.1088/1757-899X/291/1/012025

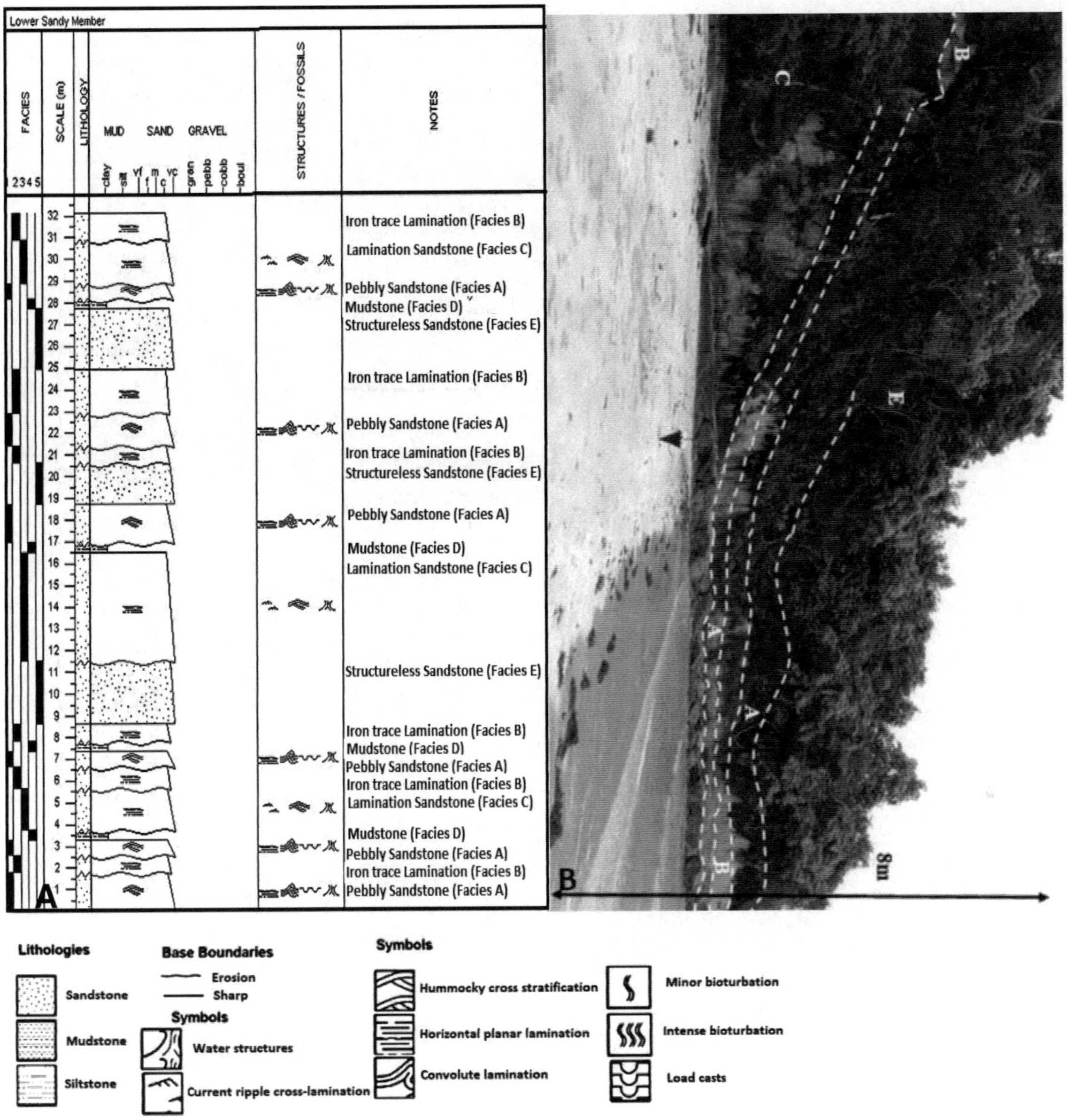

Figure 4: A) Sediment log of the Lower Sandy Member of the Kudat Formation. B) Part of facies strata of the Lower Sandy Member

These sandstones comprised very coarse to coarse graded structureless sandstone. The sands grains are well rounded and well sorted. This facies beds are being up to 1- 5 m thick. The sandstone beds are light grey to white in colour. Beds based are commonly flat. This facies in some outcrop is absent. In addition, the order of facies in some location changed and one or two facies may be absent.

Interpretation: This facies association is interpreted as deposition of concentrated density flows with erosional ability, such as high density sand-rich currents [26]. The present of convolute and flame structure which indicated as soft sediment deformation, shows the high deposition rates. The occurrence of dewatering structures in this facies also indicates that rapid deposition of sand and a short recurrence interval between depositional events [7, 27] Mud drop (mud clasts) in facies association particularly

Figure 5. A) The Pebbly sandstone consist of three section which are aggregation of pebbles (AP), two or three sets of sandstone and siltstone laminations (SL) and non-aggregation pebbly (NAP) sandstone. B) Dish structures (shown by arrows) in non- aggregation pebbly sandstone. C) Hummocky cross like in different scale in pebbly sandstone. D) Convolute structure (shown by arrows) in pebbly sandstone. E) Honeycomb (weathering structures) in Iron trace sandstone. F) Small-scale cross sets (shown by arrows) in facies C. G) Flame structure (shown by arrows) in facies C. H) Stone balls in different scale are presented in the Lower Sandy Member. These stone balls generally were arranged in a line along strike direction.

were formed by high density currents that erosion the previously deposited and after deformed the mud clasts, deposited in new sediments bed [7, 28-31]. Planner lamination form during high sediment-fallout and/or near-bed sediment concentration and generally the flows are fast and the sediment concentration is high [32]. The thin small scale cross-stratified which overlying indicated as deposition from lower energy current such as dilute tail part of flows [33].

The load structure form when dens sediments deposited over less- denser sediments which is liquefied and create Rayleigh-Taylor instability.
The presence of these dewatering and loading structures in sediments beds interpreted as rapid deposition and a short recurrence interval between depositional events. Dewatering structure indicated, insufficient time for the underlying bed to dewater before deposition of the overlying bed [7, 27]. The occurrence of hummocky cross like also indicates this facies association deposited by density flow (frictional) not tempestites currents. Hummocky cross-stratification commonly formed in tempestites but small scale of hummocky that named hummocky cross like [34] indicated to form in frictional flows [35].
In addition the occurrence of thick structureless sandstone at the top of strata also consistent with other evidence this facies association formed by long duration frictional flows [26, 36]. The present of dewatering structure also support this interpretation [26, 36, 37].

4.2. The Middle Muddy Member (Sikuati)
The Sikuati Member crops out in central part of Kudat peninsula. Most of data discussed here were collected from the 26 outcrops along main road and coast line (Figure 6). Total ~142 Meter in the area of ~ 66.4 km2 was measured in detail. The Sikuati Member of Kudat Formation was subdivided into nine facies (Table 2) and three facies association, with several to tens of meters thick of sediments according to their sedimentary properties. Thin- and medium-bedded sandstone and mudstone (MMFA1), fine sandstone commonly with planar-laminated carbonaceous and siltstone (MMFA2) and fine to coarse-grained thick basal, overlain by a thinner succession fine-grained sandstone bedding with shale or mudstone (MMFA3).
In conjunction with the broad lithologic tract, the facies could be described by stacking arrangement and geometry of the beds (two-dimensional), grain size, internal stratification structure and architecture, sandstone richness, degree of amalgamation and thickness.

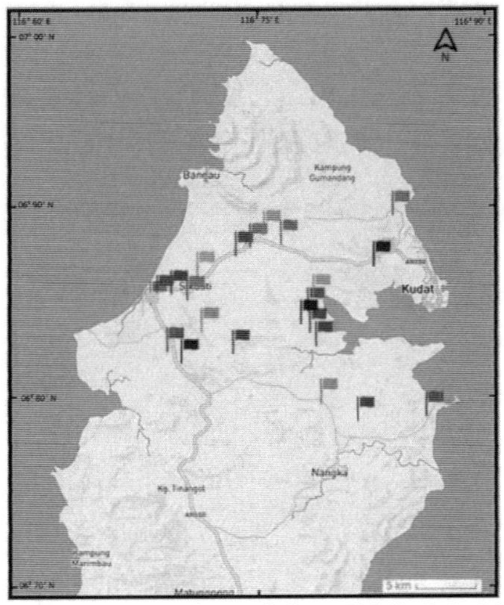

Figure 6: Outcrops location of the Middle Muddy Member

4.2.1. MMFA1: Thin- and medium-bedded sandstone and mudstone
Description: This facies association includes three facies (Figure 7) A, B and C, beginning with thin- and medium-bedded sandstone and mudstone (in some outcrops bedded with shale) (Facies A). The bed sets of facies association range from <10 cm to ~ 60 cm thick. The base of this facies consists of beds that are commonly flat and mostly sharp, but undulating contacts with shallow scour sedimentary structure.

Table 2: Facies description of Middle Muddy Member

Name	Description	Facies
Thin and medium bedded sandstone and mudstone (Facies A)	Thin and medium bedded sandstone and mudstone. Bed sets of facies association range from <10 cm to ~ 160 cm to ~ 160 cm thick.	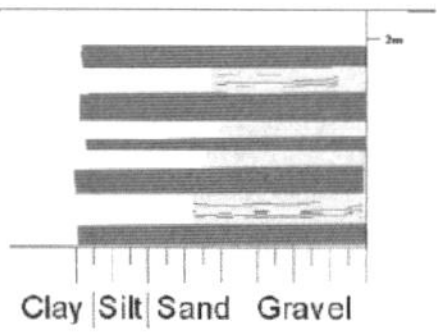
Cross-bedding Sandstone (Facies B)	Comprising only a single set of cross-bedding at the top of beds that overlain on the planar-laminated carbonaceous and mudstone.	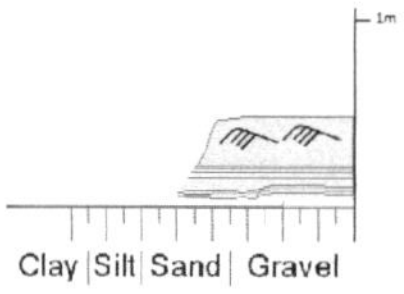
Structureless sandstone (facies C)	Fine to very-fine structureless sandstone	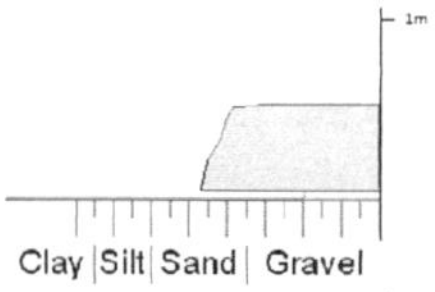
Planar-laminated sandstone (facies D)	Fine sandstone commonly with planar-laminated carbonaceous and siltstone. The planar-stratified structures are commonly syndepositionally deformed and faulted.	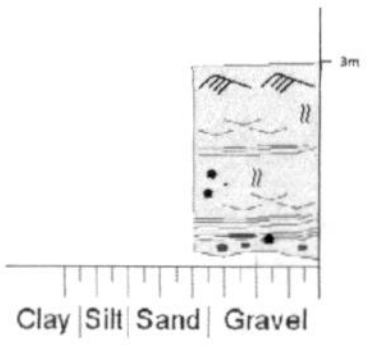
Shale (Facies E)	In some outcrops strata, vegetation took place from shale beds.	
White coarse sandstone (Facies F)	Comprising white and coarser sandstone and in his facies the beds become thinner.	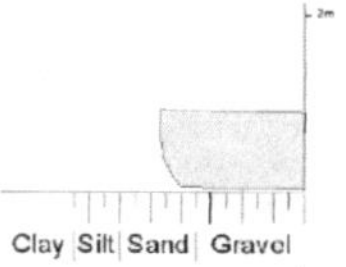
Fine to coarse-grained sandstone (facies G)	Thick fine to coarse-grained sandstone	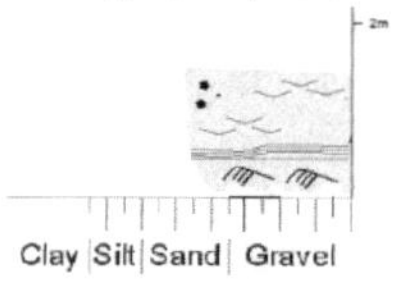
Thin sandstone and mudstone bedding (Facies H)	Comprising thin fine-grained sandstone bedding with shale or mud	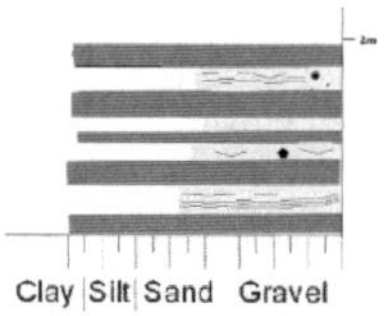
Mud drop sandstone (Facies I)	Sandstone beds comprised mud drop structures (even rare). Mud rip up structures are up to several centimeters long	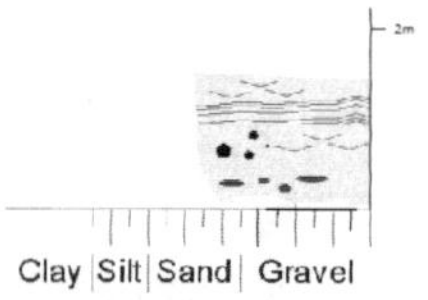

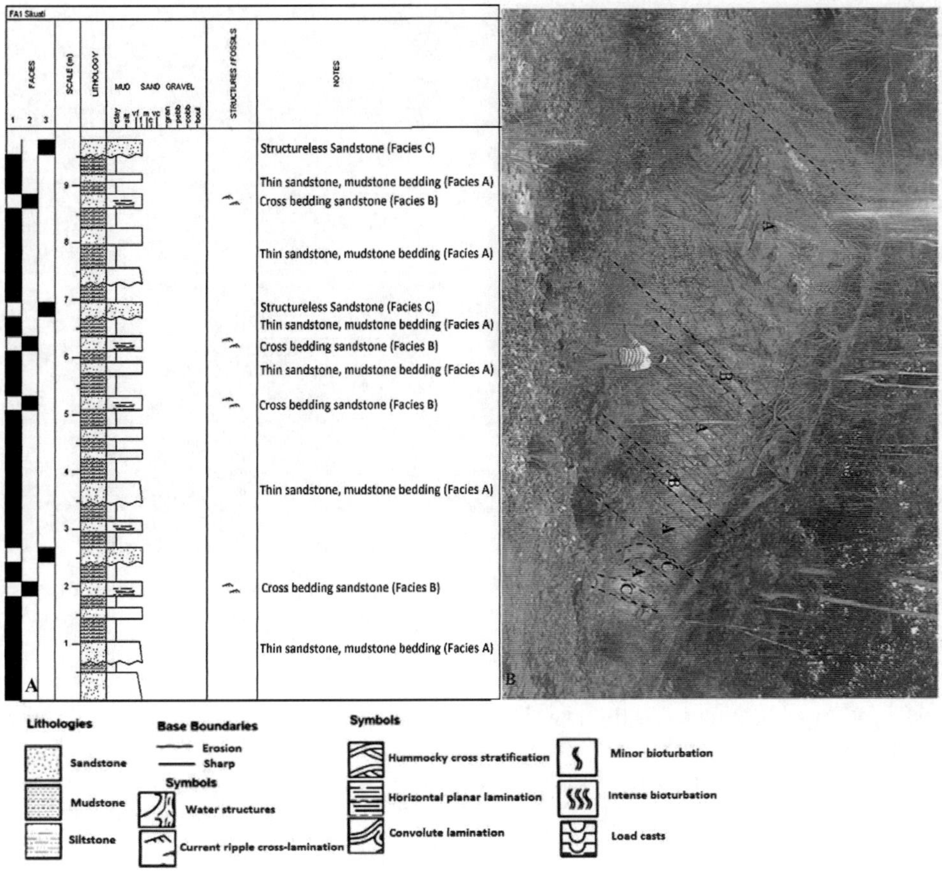

Figure 7: A) Sediment log of the Middle Muddy Member FA1 of the Kudat Formation. B) Part of facies strata of the Lower Sandy Member FA1

Sandstones are generally dark grey to light grey in colour, with several intervals up to several meters thick are conspicuously white stained. Slumps deposits (Figure 8A) are common in this facies association. Some sandstone beds (Facies B) typically comprise only a single set of cross-bedding at the top of beds. This cross-bedding overlain gradationally on planar-laminated carbonaceous and mudstone. Both planar- and cross-stratified structures are commonly deformed. Diffusely planar laminated units cap virtually most of the sandstone beds, and consist of dark carbonaceous siltstone (Figure 8B) (red colour was also observed). The thickness of planar laminated varies not only between beds, but also laterally within a single bed. The thickness of the bed varies from < 1 cm to ~ 5 cm. The third facies, facies C, is typically structure-less fine to very-fine sandstone. Thicker intervals of this facies association from few meters to tens of meters and are easily identified in the field. Although rare, some trace fossil in white sandstone have been observed at the base of some strata (Figure 8C1 & C2).
Interpretation: This facies association is characteristic of the turbidite deposition in deep marine. The shallow basal scours of the succession suggest that flow was less concentrated and/or had less energetic turbulence.
The abrupt change from the structure-less sandstone at the base of some beds to the parallel lamination and cross-bedding analogous interpret as Bouma division A, B and C in the classical turbidite [10]. In such case, as the flow continued to decelerate, parallel lamination and a thin layer of small-scale cross-stratification suggests that the flow carries lower sediment concentration and deposits it from lower energy of dilute tail part of the flows [33]. The mudstone beds also analogous to Bouma division D.

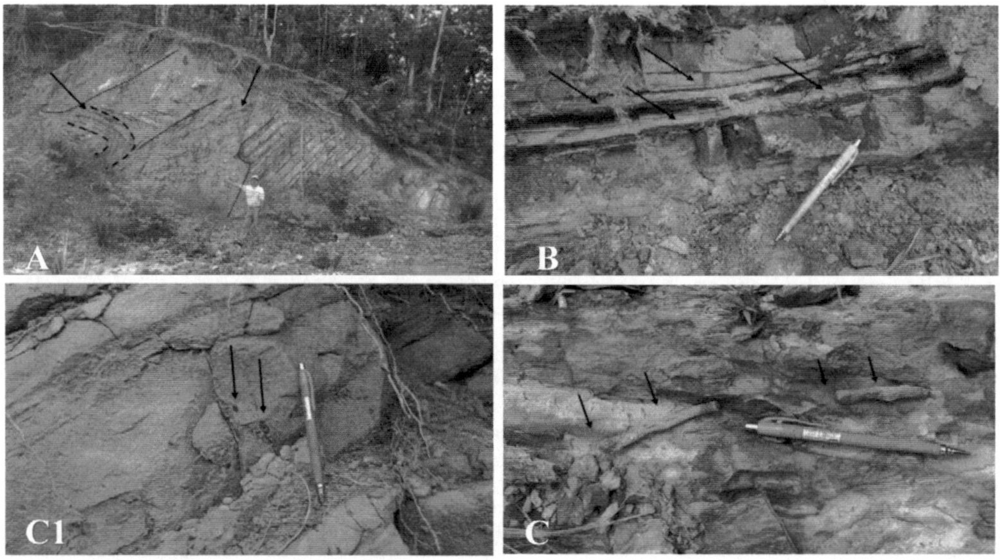

Figure 8: A) Slump feature is common in FA1. B) Planar carbonaceous lamina (shown by arrows) C1) Trace fossils (shown by arrows) in white and structureless sandstone in FA1. C2) Trace fossil (shown by arrows) in strata of FA1

This facies association is interpreted as the classic thick bedded turbidite that include 10 to 60 cm thick sandstone similar to Bouma division.

The presence of slumps within stratified blocks support this interpretation. The slumps suggested the MMFA1 is formed in continental slopes and submarine canyon heads. Slumps formed due to failure of semi-consolidated sediment mass and hydro plastic deformation could occur in most part of the sedimentary system [7, 29, 38].

4.2.2. MMFA2: Fine sandstone with planar-laminated carbonaceous siltstone

Description: The second facies association consist of three facies (Figure 9) comprising fine sandstone (Facies D) commonly with planar-laminated carbonaceous siltstone. The planar-stratified structures are commonly syndepositionally deformed and faulted. The beds are typically ~0.50 to 3 m thick and form bed-sets up to ~45 m thick. The sandstone beds are grey in colour. The bases of the sets are commonly erosional with shallow scours up (Figure 10A) to 3 to 15 cm deep and the bed contacts are typically clear and are easily demarcated.

Small number of beds contain coal pebble (Figure 10B) and bioturbation. Red-brown coloured cement, are also common in a small number of beds. Centimetre-scale dewatering structures are abundant locally (Figure 10D). The cross –bedding are common at the top of sandstone beds (Figure 10C) and distorted cross beds may indicate dewatering or soft- sediment deformation. Mudstone clasts (Figure 10D) vary from absent to abundant and in some cases, make up about 5% to 10% of the total volume in some beds. Mudstone clasts (mud rip up or chaotic mud) vary from small round clasts of several centimetres in diameter to elongated clasts up to ~10 cm thick and occur mostly near and at the bottom of the bed. These mud clasts were originated from upstream due to erosion of previously deposited and are commonly sheared and highly plastically deformed. They indicated shearing within the flow which was sufficient to deform them.

In some outcrops strata, vegetation took the place of the shale beds (Facies E), between the sandstone beds. In one of the outcrops at the top of the strata the sandstone grain size became coarser (Facies F), white colour and the beds became thinner (~0.10 – 0.9 m).

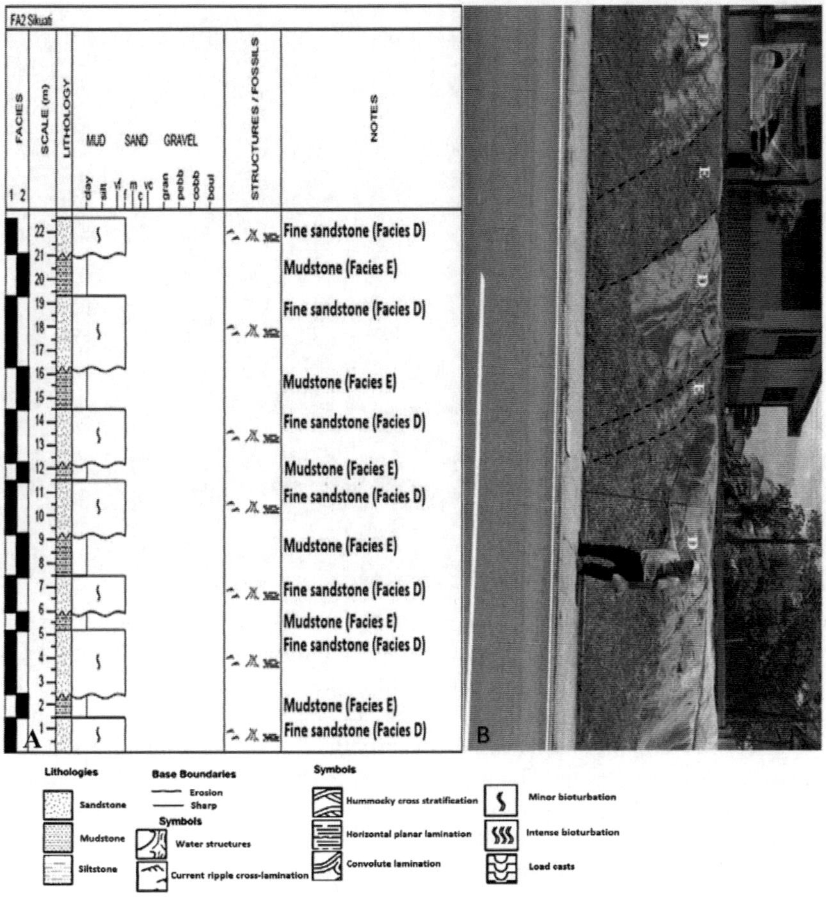

Figure 9: A) Sediment log of the Middle Muddy Member FA2 of the Kudat Formation. B) Part of facies strata of the Lower Sandy Member FA2

Interpretation: The MMFA2 is interpreted as deposits of hyper-concentrated flows with erosional ability, such as high-density sand-rich currents.The parallel lamination indicated deep bottom hyper-concentrated flows and traction currents [26, 36]. The cross-stratification part of MMFA2 has similar characteristics to the MMFA1, and this is reflected in a similar interpretation.

The occurrence of dewatering structures, deformed parallel and cross stratification in strata of MMFA2 indicates that rapid deposition of sand and a short recurrence interval between depositional events [7, 27] and an insufficient time for the underlying bed to dewater before deposition of the overlying bed. The presence of scours characterized as MMFA2 were deposited from fast-flowing energetic concentrated density flows also centimetres deep scours indicated downward energetic sweeping eddies formed shallower scours [26].

Mudstone clasts in MMFA2 particularly common in the thicker beds and were sourced from erosion of previously deposited. The high-density sand-rich currents deformed and sheared the mud clasts in deposition process [7, 28-31].

4.2.3. MMFA3: Fine to coarse-grained thick basal, overlain by a thinner succession fine-grained sandstone bedding with shale or mudstone

Description: This association include four facies and is made up of a thick basal part (~0.5 to1.5 m) (Figure 11), comprising fine to coarse-grained sandstone (Facies G), overlain abruptly by a thinner succession comprising fine-grained sandstone bedding with shale or mudstone (Facies H) with common

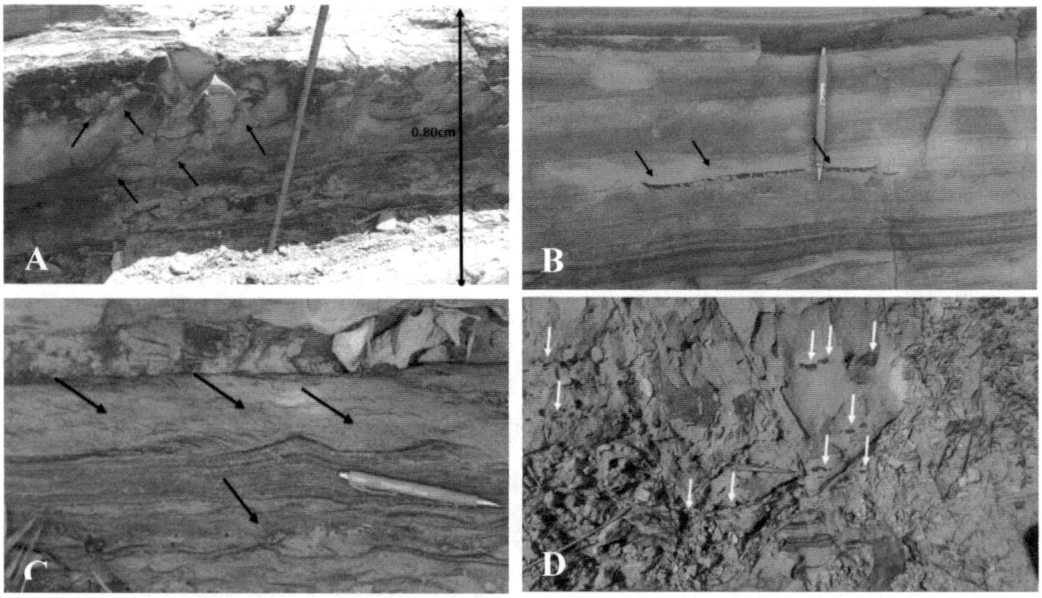

Figure 10. A) Erosional surface with shallow scours up (shown by arrows) of bed base in FA2 B) Coal pebble (shown by arrows) in strata C) Cross stratification (shown by arrows) at top of some beds D) Mud clasts (shown by arrows) in the base of strata FA2

traction sedimentary structures and the beds are ~0.10 to 0.7 m thick. Bed bases are commonly undulating contacts with shallow scours up. The sandstone beds (both, thick and thin) have a distinctly dark grey to grey colour. These beds comprise a unit of planar-laminated carbonaceous and siltstone. The thickness of planar laminated varies between beds and may vary from < 1 cm to ~ 5 cm. The climbing cross-stratified (Figure 12A) sandstone were observed in some beds overlaid by planar laminated carbonaceous, and commonly occurs as single 5-10 cm thick sets. The cross and planar stratified both are commonly syndepositionally dewatering and deformed. At the base of some sandstone beds, mud drop structures (Facies I) (even rare) are present locally. Mud rip up structures are up to several centimetres long. In a small number of beds coal pebble and loading structures (Figure 12B) although uncommon, occurred.

At the top of strata in two outcrops the thick massive sandstones (Facies D) have been observed interbedded by thin shale beds (~0.1 to 0.2 m). These sandstone beds are being up to about 1-3 m thick and comprised coarse graded structure-less sandstone.

Interpretation: The shallow scours along bed contacts has similar characteristics as strata in FA1, and this is reflected in a similar depositional interpretation.

The planar lamination is interpreted, to be the result of high sediment-fallout and/or near-bed sediment concentration and the flows were faster and/or the sediment concentration high [32]. The thin overlying small-scale cross-stratified indicate deposition of new sediment sourced from the low energy, dilute tail part of flows and flow speeds in the dilute tail part of the current are maintained for periods of time. (analogous to the Bouma Tc division [33]).

The present of dewater and load structures are most common in strata of this facies association. Dewater structures are formed by escaping pore fluids when a significantly high pores water pressure gradient exists. This gradient can be achieved by rapid loading of loosely packed, water-saturated sediment [7] that in turn can create Rayleigh-Taylor instability at the sediment interface, resulting in load structures. The presence of these dewatering and loading structures in sandstone beds characterized rapid deposition of sand and a short recurrence interval between depositional events and an insufficient time for the underlying bed to dewater before deposition of the overlying bed [7, 27].

IOP Conf. Series: Materials Science and Engineering **291** (2017) 012025 doi:10.1088/1757-899X/291/1/012025

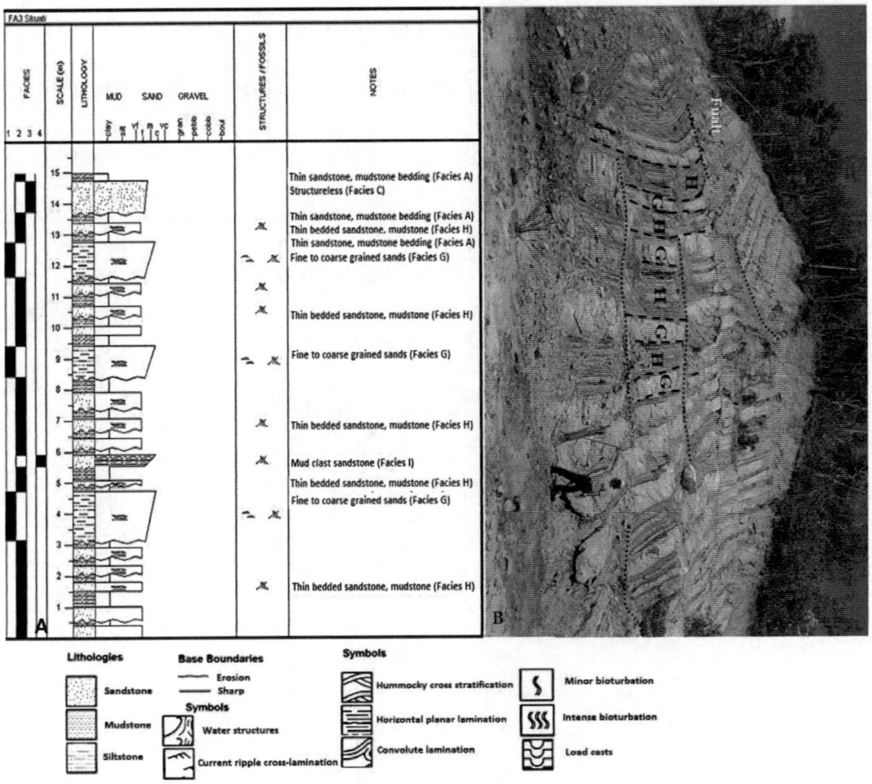

Figure 11: A) Sediment log of the Middle Muddy Member FA3 of the Kudat Formation. B) Part of facies strata of the Lower Sandy Member FA3

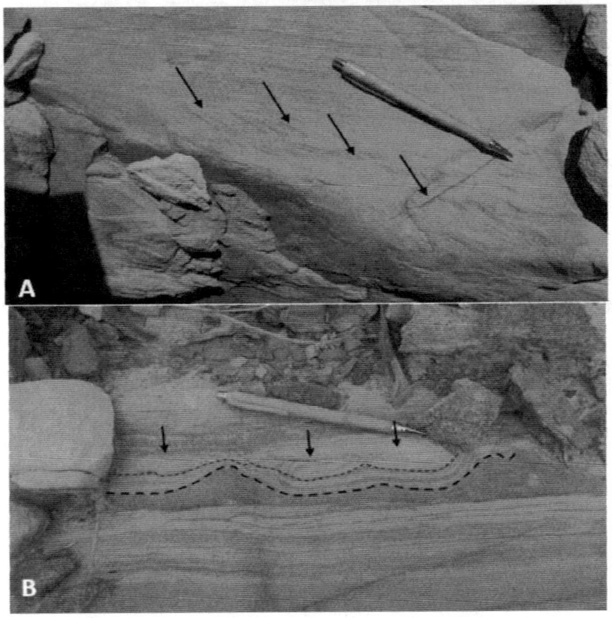

Figure 12: A) Climbing cross stratification in FA3(shown by arrows). B) Uncommon load structure in some strata of FA3

Mudstone clasts are also common in strata of this facies association. These clasts along bottom part of bed has similar characteristics as FA2.

Massive beds are thought to have been deposited from the base of the flow by gradual although rapid aggradation [7].

4.3. The Upper Sandy Member (Gomontong)

The Upper Sandy Member crops out in Southern part of Kudat peninsula. All data were collected from the 13 outcrops along main road and coast line (Figure 13). Total ~500 meter in the area of ~ 81 km2 was measured in detail. Field work strata of the Upper Sandy Member investigation identified two associated facies. Each facies and lithology have been study in detail and lithological description, thin section analysis was used to characterize each facies (Table 3).

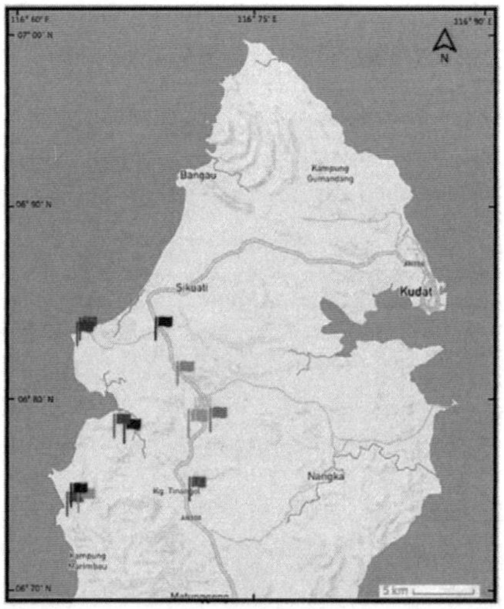

Figure 13: A) Outcrops location of the Upper Sandy Member

4.3.1. USFA1: The upper sandy fine to very fine grain sandstone

Description: The facies association include three facies (Figure 14) and is made up of a facies A, that comprising fine to very fine grain sandstone with undulated carbonaceous laminations. The sandstone beds thickness range from 0.2 to 1.70 m and bed base are undulating with shallow scours up. The undulated carbonaceous laminations thickness varies between beds and also laterally along a single bed and ranges from 2 to 50 mm (Figure 15A). In most of beds some coal pebbles have been observed (Figure 15B). Sandstone beds are dark grey. This facies generally interbedded with mudstone beds which mudstone thickness are 10-25 cm and syndepositionaly deformed and faulted. This facies overlain abruptly by structureless sandstone, facies B. This facies comprised fine to medium graded structureless sandstone (in one outcrops, coarse graded sands also have been observed). These sandstone beds are being up to about 0.80-7.70 m thick. This sandstone beds are light grey to dark grey and beds based are commonly flat with no or shallow scours up. At the top of strata, coarse grain sandstone with straight to undulated carbonaceous laminations (Facies C) over lied facie A and B. Facies C consists of medium grain sandstone with straight to undulated carbonaceous laminations. The beds based are commonly undulating. The sand grains are sub rounded to rounded and moderate sorted and the sandstone are distinctly light grey in this facies. The planner lamination thickness ranges from 2 to ~5 cm. The sandstone beds thickness ranges from 0.7 to 1.6 m and interbedded with shale or same facies.

Interpretation: This facies association like the only facies association of Lower Sandy Member interpreted as deposition of concentrated density flows. All sedimentary structure consistent with this interpretation.

IOP Conf. Series: Materials Science and Engineering **291** (2017) 012025 doi:10.1088/1757-899X/291/1/012025

Mud drop (mud clasts) in facies association particularly were formed by high density currents that erosion the previously deposited and after deformed the mud clasts, deposited in new sediments bed [7, 28-31]. In addition the planner and undulating lamination in this facies generally formed during high velocity and/or high sediment concentration flows [32].

Table 3: Facies description of Upper Sandy Member

Name	Description	Facies
Fine to very fine grain sandstone (Facies A)	Fine to very fine grain sandstone with undulated carbonaceous laminations.	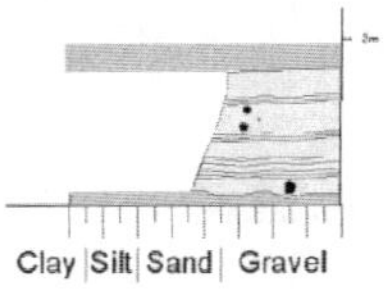
Fine to medium graded structureless sandstone (Facies B)	Comprising fine to medium graded structureless sandstone.	
Coarse lamination sandstone (Facies C)	Consisting of medium grain sandstone with straight to undulated carbonaceous laminations. The beds based are commonly undulating.	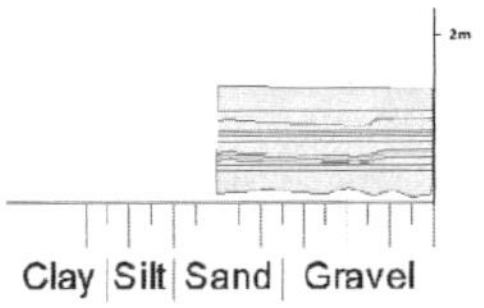
Thin bedded sandstone and mudstone (Facies D)	Comprising thin bedded sandstone and mudstone, which mudstone is dominated. Sandstones are generally structureless.	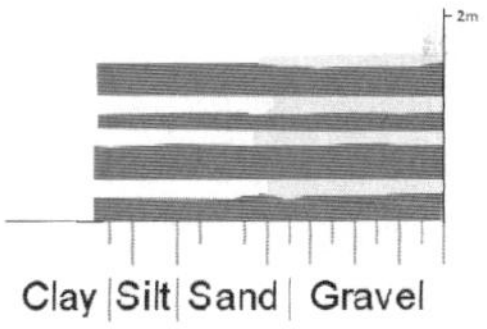

4.3.2. USFA2: The upper sandy thin bedded sandstone and mudstone

Description: The second facies association consist two facies. It comprises thin bedded sandstone and mudstone, facies D and in some location facies A (Figure 16). Facies D comprises thin bedded sandstone and mudstone, which mudstone is dominated. Bed sets of facies range from <5 cm to ~ 15 cm thick. Bed bases are commonly undulating to moderately flat. Sand grains of sandstone are fine to very fine in size. Sandstones are generally structureless and light grey to white in colour. This facies is mainly same as facies A in FA1 in Sikuati Member. Thicker interval of this facies is easily identified in the field. Some sandstone beds (facies A) typically comprise fine to very fine grain sandstone with undulated carbonaceous laminations.

Interpretation: This facies association is characteristic of the turbidite (hyperpycnal turbidites) deposition such as FA1 in Sikuati Member.

The shallow basal scours of the succession suggest that flows were less concentrated and/or had less energetic turbulence.

The abrupt change from the structureless sandstone at the base of some beds to the parallel lamination and cross-bedding analogous interpret as Bouma division A, B and C in classical turbidite [10]. The flow continued to decelerate parallel lamination and a thin layer of small-scale cross-stratification, that suggested the flow have had lower sediment concentration and deposited from lower energy, dilute tail part of flows [33].The mud beds also analogous to Bouma division D.

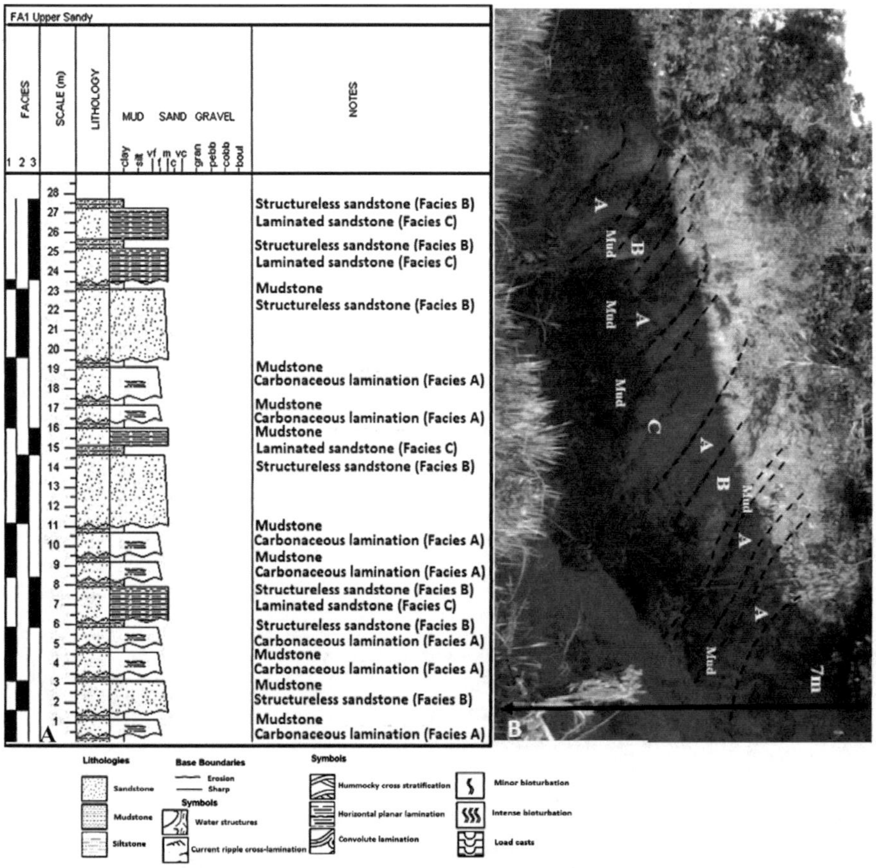

Figure 14. A) Sediment log of the Upper Sandy Member FA1of Kudat Formation B) Part of facies strata of Upper Sandy Member FA1

Figure 15. A) Carbonaceous laminations in USFA1 is Undulated and showed varies thickness B) Carbonaceous laminations and Coal pebbles in USFA1(shown by arrows)

IOP Conf. Series: Materials Science and Engineering **291** (2017) 012025 doi:10.1088/1757-899X/291/1/012025

This facies association interpreted as the classic thick bedded turbidite that include 10 to100 cm thick sandstone and tend to be composed of Bouma's division.
As a final point the presence of slumps with stratified blocks support this interpretation. The occurrence of slumps suggested the FA1 is formed in continental slopes and submarine canyon heads. Slumps formed due to failure of semi-consolidated sediment mass and hydro plastic deformation and it occurred in most part of the sedimentary system. [7, 29, 38].

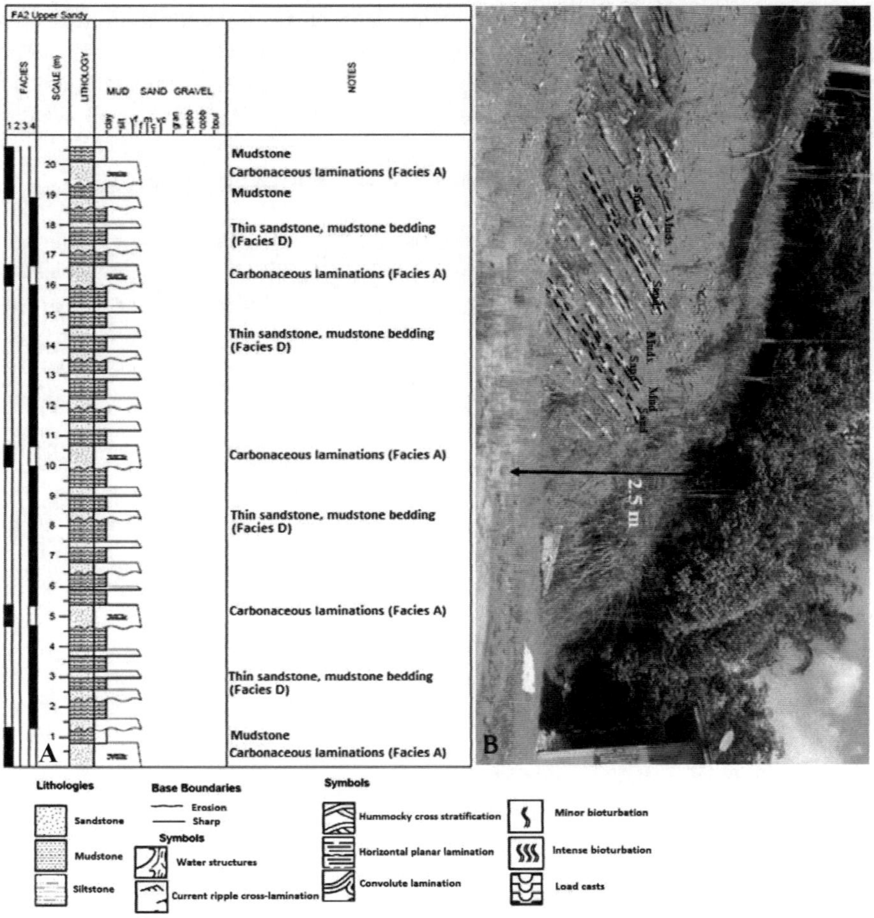

Figure 16. A) Sediment log of the Upper Sandy Member FA2 of Kudat Formation B) Part of facies strata of Upper Sandy Member FA2

5. Architectural Elements

To characterize architecture elements and litho-stratigraphic framework, detailed sedimentary facies association, depositional architecture and environment context were identified.
In this study area six architectural elements were characterized: channel fill, inner external levee, lobes sheet, overbank terrace

5.1. LSFA1: Channel Fill

The association of irregular, pebbly sandstone, iron trace and lamination sandstone and mudstone, structure-less sandstone in FA1, is interpreted as channel fill (Table 4).

Channel comprised of shallow erosional base filled with pebbly sandstone that representing deposition from low-density currents. The occurrence of mud clast at the base of some beds shows initial fill eroded from the channel bottoms and sides.

The presents of lamination sandstone with small scale cross bedding in channel fill also indicated the sediment deposited by low density current approximately at distal part of channel complex [39-41]. In contrast the present of structure-less sandstone shows this facies formed from high density current with high sediment concentration [39-41] and also it indicated the velocity and density of flows were fluctuated. The considerable thickness of structure-less sandstone in some outcrops (up to 5 m) representing these beds deposited by long lived and generally steady flows.

Masalimova and his colleagues [41] argue that if no obvious fining upward trends characterized in the sediments beds is because of limited range of grain size.

Table 4: Sedimentary facies and element architecture of LSFA1

Facies Association	Facies	Lithology	Primary structures	Ichnofacies	Architecture Elements
Thick and medium bedded sandstone and mudstone (LSFA1)	A	Thin and medium bedded pebbly sandstone	Laminated carbonaceous coal pebble, mud clasts, dewatering structures, cross –bedding, flame and convolute structure	Non	Channel fill: Sediments deposited from high to low density currents with different value of sediment concentration in channel
	B	Iron trace sandstone	Laminated carbonaceous	Non	
	C	Lamination sandstone	Laminated carbonaceous coal pebble, hummocky like, dewatering structures, cross –bedding, flame and convolute structure	Non	
	D	Mudstone	Laminated carbonaceous	Non	
	E	Structureless sandstone		Non	

5.2. MMFA1: Inner External Levee

The architectural element A is characterized by regular interval of thin bedded sandstone and shale and/or mudstone in FA1, and interpreted as inner external levee. This element is formed when distal overbank turbidity flows spill out from their confinement of channel belt. The deposition take place outside channel belts and basically has significant lateral bed thickness variation and deformation [42]. The deformation of AEA at inner external levee, architecturally characterized as a result of slumping directed to the channel belt. The thicker sandstone beds (>40 cm) are characterized as crevasse splays when flows cut the levee and deposit sand. In the external levee, thin beds are deposited perpendicular to the channel belt deposition. Further the slope gradient is steeper into the channel belt than the slope dis away from channel. The inner external levees located between outer portion of external and internal levees and they are normally formed during regressive time [42-44] (Table 5).

5.3. MMFA2: Lobes Sheet

The architectural element B is characterized by consisting of interbedded fine massive thick sandstone (up to 3m), it interbedding with shale and it is characteristic of lobes deposition. The individual lobe is showing more tabular or sheet like than lensoid and/or lobate particularly in the large lobes and mature channel system [7].

Table 5: Sedimentary facies and element architecture of MMFA1

Facies Association	Facies	Lithology	Primary structures	Ichnofacies	Architecture Elements
Thin- and medium-bedded sandstone and mudstone (MMFA1)	A	Several meters thick, dark grey to light grey sandstone.	Laminated carbonaceous	Non	Inner external levee: Distal overbank turbidity flows spill out from their confinement of channel belt
	B	Sandstone with cross-bedding over planar lamination carbonaceous	Cross-bedding and planar-laminated carbonaceous	Non	
	C	Structure-less fine to very-fine sandstone		Trace fossil	

This element formed in lob complex, as downslope sandy hyper-concentrated flows develop lob like features. Massive sandstones are inferred to have been deposited by rapid sedimentation from high-density currents. Shallow and broad dishes dewatering structures indicated as result of fast deposition rate and effective weight of the overlying sediments and also fluidization process [7].
This element deposited in fully mature channel system is feeding stacked of lobe complexes associated with levees. The shale clasts found at the base of the bed are the result of high energy erosive rip-up of the underlying shale and indicate high energy erosion [7] (Table 6).

Table 6: Sedimentary facies and element architecture of MMFA2

Facies Association	Facies	Lithology	Primary structures	Ichnofacies	Architecture Elements
Fine sandstone commonly with planar-laminated carbonaceous and siltstone (MMFA2)	D	Grey, fine sandstone with planar-laminated carbonaceous	Laminated carbonaceous coal pebble, dewatering structures, cross – bedding, Mudstone clasts	Trace fossil	Lobes sheet: Deposited in fully mature channel system feeding stacked lobe complexes and associated with levees
	E	Siltstone	Laminated carbonaceous, dewatering structures, cross –bedding	Non	
	F	White , coarse and thin bedded sandstone		Non	

5.4. MMFA3: Overbank Terrace

The architectural element C is characterized by the association of irregular, thick-bedded sandstones interbedded with thin sandstones and mudstone of FA3, this is typically an overbank terrace where the deposition of sediments is on a gentle gradient margins of the channel belt. The overbank terrace is formed by rapid suspension sedimentation from collapsing sand rich, high density currents or/and by unconfined sandy flow and also off-axis turbidity currents. This deposition is common on low stand trend and in low and middle stage of channel complex sets [45-47].
In FA3 the occasional of thin sandstone and mudstone beds intervals suggests that the sediments are coming from suspension during low energic flow. On the other hand, the thick sandstone beds are

interpreted to be coming from higher energy currents or/and unconfined flows resulting in transporting greater material and thicker sandstone beds (Table 7).

5.5. USFA1: Channel Fill

The association of irregular, pebbly sandstone, iron trace and lamination sandstone and mudstone, structure-less sandstone in FA1, is interpreted as channel fill (Table 8).

Channel comprised of shallow erosional base filled with pebbly sandstone that representing deposition from low-density currents. The occurrence of mud clast at the base of some beds shows initial fill eroded from the channel bottoms and sides.

The presents of lamination sandstone with small scale cross bedding in channel fill also indicated the sediment deposited by low density current approximately at distal part of channel complex [39-41]. In contrast the present of structure-less sandstone shows this facies formed from high density current with high sediment concentration [39-41] and also it indicated the velocity and density of flows were fluctuated. The considerable thickness of structure-less sandstone in some outcrops (up to 5 m) representing these beds deposited by long lived and generally steady flows.

That was argued if no obvious fining upward trends characterized in the sediments beds is because of limited range of grain size [41].

Table 7: Sedimentary facies and element architecture of MMFA3

Facies Association	Facies	Lithology	Primary structures	Ichnofacies	Architecture Elements
Fine to coarse-grained thick basal, overlain by a thinner succession fine-grained sandstone bedding with shale or mud (MMFA3)	FG	Fine to coarse-grained sandstone	Planar-laminated carbonaceous and siltstone, climbing cross-stratified, coal pebble and loading structures	Non	Overbank terrace: Forms by rapid suspension sedimentation from collapsing sand rich, high density currents or/and by unconfined sandy turbidity flow
	FH	Fine-grained sandstone bedding with shale or mud	Planar-laminated carbonaceous and siltstone, climbing cross-stratified, coal pebble	Non	
	FI	Sandstone beds with mud drop structures	Mud drop	Non	
	FD	Thick massive sandstones interbedded by thin shale beds		Non	

5.6. USFA2: Lob Complex

The intervals of two facies association of Upper Sandy Member interpreted as lob complex (Table 9). This lob complex consists of very fine grain sandstone with undulated carbonaceous laminations, fine to medium graded structureless sandstone, medium grain sandstone with straight to undulated carbonaceous laminations and thin bedded sandstone and mudstone, which mudstone is dominated.

In terms of rapid collapsing high density current (turbidite) in distal lob the structureless sandstone were deposited. In proximal lobes deposits generally, dune like structure is observed in contrast of distal lobes which planner lamination and or thin cross lamination structure is dominated. During high density currents, when the flows were steady the dune like and bed load transport occurred. Further downslope, when current decline and flow collapse accelerated, structureless beds formed [7, 39-41]. Masalimova [41] argue that coarser structureless sands form by high suspended-sediment fallout in more proximal lobe.

Table 8: Sedimentary facies and element architecture of USFA1

Facies Association	Facies	Lithology	Primary structures	Ichnofacies	Architecture Elements
FA1: The Upper Sandy fine to very fine grain sandstone (USFA1)	A	Fine to very fine grain sandstone	carbonaceous laminated coal pebble	Non	ArC: Channel fill: Sediments deposited from high to low density currents with different value of sediment concentration in channel
	B	fine to medium graded structureless sandstone		Non	
	C	Coarse lamination sandstone	undulated carbonaceous laminations	Non	

When high density flows lost the strength, sand deposition occurred within channel, and small amount of very fine and mud reaching to the lobe complex and facies B formed [40, 41].
In conclusion the Upper Sandy Member indicated as lobe distal complex of comprising thinning-upward lobe cycles of structureless and cross-stratified thick-to thin-bedded sandstone separated by mudstone.

Table 9: Sedimentary facies and element architecture of USFA

Facies Association	Facies	Lithology	Primary structures	Ichnofacies	Architecture Elements
The Upper Sandy thin bedded sandstone and mudstone (USFA2)	D	Fine to coarse-grained sandstone	Planar-laminated carbonaceous and siltstone, climbing cross-stratified, coal pebble and loading structures	Non	Overbank terrace: Forms by rapid suspension sedimentation from collapsing sand rich, high density currents or/and by unconfined sandy turbidity flow
	A	Fine-grained sandstone bedding with shale or mud	Planar-laminated carbonaceous and siltstone, climbing cross-stratified, coal pebble	Non	

6. Discussion and Depositional Environment

During Miocene the region which include the area of this study is undergoing complex tectonic activity (both extensional and compressional stresses) [48] and the presence of mélange deposition in Kudat Peninsula is evidence of this complexity [21, 22, 49, 50]. Besides that, this tectonic activity can be caused of gravity flow, slumping, and other turbulent currents.

It is demonstrated that the Kudat Formation was deposited in Lower Miocene intimately with the Crocker Formation in shallow to deep marine environment [18, 22, 50]. Tongkul [21] believed the Crocker and Kudat Formation are both formed in deep marine environment. However, the Kudat Formation also shows shallow marine properties evidence, and he suggested this formation was deposited in shelf slope.

Cronin [9] demonstrates that deep water canyon system has three different characteristic geographic regions and stages, those are erosional area, erosional and depositional area and depositional area in order through shelf slope to sea floor.

On the other hand, it suggested that in general the channels form in three phases: erosion and bypass, channel-fill and over-bank deposits, and abandonment. The initial phase makes a typical channel U shape by erosion and bypass processes. In the next phase, due to decreasing in transport energy the channel- fill and over-banks deposits. At the base of the channel in this phase the poorly sorted pebbly sandstones and mudstones are deposited and most of the sandy material continues to by-pass this depression towards the sea basin.With decreasing energy, the sand is deposited within the channel and latter with continued in decreasing energy the suspension sediments are deposited. The last phase occurs when numerous turbidities with suspension of fine-grained sand and mud are dominated [13]. It was argued that this phase can follow Cronin [9] stages where the first phase is characterized by erosional stage, and basal erosion of old sediments. The following phased shows a mixed erosion and deposition stage. During these processes of channel development, the slumping may occur in this stage. The last phase is characterized by deposition with transition between sand and silts followed by mud deposition [13].

A shelf to deep marine turbidite and hyper-concentrated flow depositional system is best described of the depositional scenario for the Sikuati Member. These sediments are interpreted to be formed during the second phase of channel evolution system and mostly deposited in erosion and deposition area of Cronin model. The presence of slumping and hyper-concentrated sediments indicate deposition occurred in the second stage [9, 13].

7. Conclusion

Lithofacies and sedimentary structures have provided useful data for interpret the depositional environments and modelling of the three members of Kudat Formation.

The Lower Sandy Member including one facies association (LSFA1) that including five facies. The USMFA1 is thick and medium bedded sandstone and mudstone that interpreted as channel fill deposition.

The Lower to Middle Miocene Middle Muddy Member consists of nine sedimentary facies which can be grouped into three facies association: MMFA1 is characterized by thin- and medium-bedded sandstone and mudstone that was deposited as the inner external levee. MMFA2 is typically fine sandstone with planar-laminated carbonaceous siltstone that can be interpreted as lobes sheet. MMFA3 indicated fine to coarse-grained thick sandstone overlain by a thin fine-grained sandstone bedding with shale or mudstone and this association is identified as overbank terrace.

The Upper Sandy Member is divided into two facies association with four facies. The UPFA1 is characterized as fine to very fine grain sandstone that interpreted as channel fill. Sediments deposited from high to low density currents with different value of sediment concentration in the channel. The USFA2 including thin bedded sandstone and mudstone and characterized as over bank trace which forms by rapid suspension sedimentation from collapsing sand rich, high density currents or/and by unconfined sandy turbidity flow.

In conclusion the Kudat Formation was deposited in shelf to deep marine environment as channel system and fan turbidite depositional system.

ACKNOWLEDGMENT
I would like to thank University Teknologi PETRONAS for supporting my study in this university and make it possible to produce this paper. I wish to thank the Mineral and Geoscience Department of Malaysia, Sabah for their support and allowing access to the reports.

REFERENCES

[1]. Weimer, P., R.M. Slatt, and R. Bouroullec, *Introduction to the petroleum geology of deepwater settings.* 2007.
[2]. Reading, H.G. and M. Richards, *Turbidite systems in deep-water basin margins classified by grain size and feeder system.* AAPG bulletin, 1994. **78**(5): p. 792-822.
[3]. Galloway, W.E., *Siliciclastic slope and base-of-slope depositional systems: component facies, stratigraphic architecture, and classification.* AAPG bulletin, 1998. **82**(4): p. 569-595.
[4]. Mayall, M. and I. Stewart. *The architecture of turbidite slope channels.* in *Deep-Water Reservoirs of the World: SEPM, Gulf Coast Section, 20th Annual Research Conference*. 2000. SEPM.
[5]. Morris, W., et al. *Reservoir architecture of deepwater sandstones: examples from the Skoorsteenberg Formation, Tanqua Karoo Sub-basin, South Africa.* in *Deep-water reservoirs of the world: Gulf Coast Section SEPM Foundation, Twentieth Annual Research Conference*. 2000.
[6]. Morris, W.R. and W.R. Normark, *Sedimentologic and geometric criteria for comparing modern and ancient sandy turbidite elements.* Deep-water reservoirs of the world (CD-ROM): Gulf Coast Section SEPM, 2000: p. 606-623.
[7]. Stow, D.A. and M. Mayall, *Deep-water sedimentary systems: new models for the 21st century.* Marine and Petroleum Geology, 2000. **17**(2): p. 125-135.
[8]. Sullivan, M., et al. *Architectural analysis of deep-water outcrops: Implications for exploration and development of the Diana sub-basin, western Gulf of Mexico*. in *Deep-water reservoirs of the world: Gulf Coast Section SEPM Foundation 20th Annual Research Conference*. 2000. SEPM.
[9]. Cronin, B.T., et al., *Morphology, evolution and fill: implications for sand and mud distribution in filling deep-water canyons and slope channel complexes.* Sedimentary Geology, 2005. **179**(1): p. 71-97.
[10]. Posamentier, H.W. and R.G. Walker, *Facies models revisited.* Society for Sedimentary Geology (SEPM), 2006.
[11]. Kane, I.A., et al., *Anatomy of a submarine channel–levee: an example from Upper Cretaceous slope sediments, Rosario Formation, Baja California, Mexico.* Marine and Petroleum Geology, 2007. **24**(6): p. 540-563.
[12]. Hubbard, S.M., B.W. Romans, and S.A. Graham, *Deep-water foreland basin deposits of the Cerro Toro Formation, Magallanes basin, Chile: Architectural elements of a sinuous basin axial channel belt.* Sedimentology, 2008. **55**(5): p. 1333-1359.
[13]. Labourdette, R., P. Crumeyrolle, and E. Remacha, *Characterisation of dynamic flow patterns in turbidite reservoirs using 3D outcrop analogues: Example of the Eocene Morillo turbidite system (south-central Pyrenees, Spain).* Marine and Petroleum Geology, 2008. **25**(3): p. 255-270.
[14]. Romans, B.W., et al., *Evolution of deep-water stratigraphic architecture, Magallanes Basin, Chile.* Marine and Petroleum Geology, 2011. **28**(3): p. 612-628.
[15]. Macauley, R.V. and S.M. Hubbard, *Slope channel sedimentary processes and stratigraphic stacking, Cretaceous Tres Pasos Formation slope system, Chilean Patagonia.* Marine and Petroleum Geology, 2013. **41**: p. 146-162.
[16]. Browne, G.H., et al., *Grain-size characteristics for distinguishing basin floor fan and slope fan depositional settings: Outcrop and subsurface examples from the late Miocene Mount*

Messenger Formation, New Zealand. New Zealand Journal of Geology and Geophysics, 2005. **48**(2): p. 213-227.

[17]. Muda, J., *Oil seepages at Kampung Minyak, Kudat Peninsula, Northern Sabah: Potential for geotourism development.* Bulletin of the Geological Society of Malaysia, 2010. **56**: p. 49-52.

[18]. Tongkul, F., *The geology of Northern Sabah, Malaysia: its relationship to the opening of the South China Sea Basin.* Tectonophysics, 1994. **235**(1-2): p. 131-147.

[19]. Liechti, P., F.W. Roe, and N.S. Haile, *The Geology of Sarawak, Brunei and the Western Part of North Borneo: text*. Vol. 1. 1960: Government Printing Office.

[20]. Tongkul, F., *Tectonic evolution of Sabah, Malaysia.* Journal of Southeast Asian Earth Sciences, 1991. **6**(3-4): p. 395-405.

[21]. Tongkul, F., *Structural Style of Kudat Peninsula, Sabah (Paper 35).* 2003.

[22]. Stephens, E.A., *The geology and mineral resources of the Kota Belud and Kudat area, North Borneo*. 1956: US Government Printing Office.

[23]. Wilson, R.A. and N. Wong, *The geology and mineral resources of the Labuan and Padas Valley area, Sabah, Malaysia*. 1964: Geological Survey, Borneo Region.

[24]. Tongkul, F., *The structural style of Lower Miocene sedimentary rocks, Kudat Peninsula, Sabah.* Geological Society of Malaysia, Bulletin, 2006. **49**: p. 119-124.

[25]. Pemberton, S.G., *Trace fossil facies models: environmental and allostratigraphic significance.* Facies Model: Response to Sea Level Change, 1992: p. 47-72.

[26]. Mulder, T. and J. Alexander, *The physical character of subaqueous sedimentary density flows and their deposits.* Sedimentology, 2001a. **48**(2): p. 269-299.

[27]. Lowe, D.R., *Sediment gravity flows: II Depositional models with special reference to the deposits of high-density turbidity currents.* Journal of Sedimentary Research, 1982. **52**(1).

[28]. Postma, G., W. Nemec, and K.L. Kleinspehn, *Large floating clasts in turbidites: a mechanism for their emplacement.* Sedimentary geology, 1988. **58**(1): p. 47-61.

[29]. Shanmugam, G., *50 years of the turbidite paradigm (1950s—1990s): deep-water processes and facies models—a critical perspective.* Marine and petroleum Geology, 2000. **17**(2): p. 285-342.

[30]. Johnson, S., et al., *Anatomy of basin floor to slope turbidite systems, Tanqua Karoo, South Africa: sedimentology, sequence stratigraphy and implications for subsurface prediction.* Sedimentology, 2001. **48**: p. 987-1023.

[31]. Saito, T. and M. Ito, *Deposition of sheet-like turbidite packets and migration of channel-overbank systems on a sandy submarine fan: an example from the Late Miocene–Early Pliocene forearc basin, Boso Peninsula, Japan.* Sedimentary Geology, 2002. **149**(4): p. 265-277.

[32]. Arnott, R., *Turbidites, and the Case of the Missing Dunes.* Journal of Sedimentary Research, 2012. **82**(6): p. 379-384.

[33]. Bouma, A.H., *Sedimentology of some flysch deposits: a graphic approach to facies interpretation*. 1962: Elsevier Pub. Co.

[34]. Tinterri, R., *Combined flow sedimentary structures and the genetic link between sigmoidal-and hummocky-cross stratification.* GeoActa, 2011. **10**(4).

[35]. Einsele, G. and A. Seilacher, *Distinction of tempestites and turbidites. In: Einsele, G., Ricken, W., Seilacher, A. (Eds.), Cycles and Events in Stratigraphy.* Springer-Verlag, Berlin, Heidelberg, New York,, 1991: p. 377–382.

[36]. Mulder, T. and J. Alexander, *Abrupt change in slope causes variation in the deposit thickness of concentrated particle-driven density currents.* Marine Geology, 2001b. **175**(1): p. 221-235.

[37]. Lowe, D.R. and M. Guy, *Slurry-flow deposits in the Britannia Formation (Lower Cretaceous), North Sea: a new perspective on the turbidity current and debris flow problem.* Sedimentology, 2000. **47**(1): p. 31-70.

[38]. Mohrig, D., et al., *Hydroplaning of subaqueous debris flows.* Geological Society of America Bulletin, 1998. **110**(3): p. 387-394.

[39]. Pickering, K.T., R.N. Hiscott, and F.J. Hein, *Deep-marine environments: Clastic sedimentation and tectonics*. 1989: Allen & Unwin Australia.

IOP Conf. Series: Materials Science and Engineering **291** (2017) 012025 doi:10.1088/1757-899X/291/1/012025

[40]. Carr, M. and M. Gardner, *Portrait of a basin-floor fan for sandy deepwater systems, Permian Lower Brushy Canyon Formation, West Texas.* SPECIAL PUBLICATION-SEPM, 2000. **68**: p. 215-232.

[41]. Masalimova, L.U., et al., *Outcrop characterization of a submarine channel-lobe complex: The Lower Mount Messenger Formation, Taranaki Basin, New Zealand.* Marine and Petroleum Geology, 2016. **71**: p. 360-390.

[42]. Kane, I.A. and D.M. Hodgson, *Sedimentological criteria to differentiate submarine channel levee subenvironments: exhumed examples from the Rosario Fm.(Upper Cretaceous) of Baja California, Mexico, and the Fort Brown Fm.(Permian), Karoo Basin, S. Africa.* Marine and Petroleum Geology, 2011. **28**(3): p. 807-823.

[43]. Deptuck, M.E., et al., *Architecture and evolution of upper fan channel-belts on the Niger Delta slope and in the Arabian Sea.* Marine and Petroleum Geology, 2003. **20**(6): p. 649-676.

[44]. Kane, I.A., et al., *Architecture of a coarse-grained channel–levée system: the Rosario Formation, Baja California, Mexico.* Sedimentology, 2009. **56**(7): p. 2207-2234.

[45]. Navarro, L., Z. Khan, and R.W.C. Arnott, *Depositional architecture and evolution of a deep-marine channel–levee complex: Isaac Formation (Windermere Supergroup), Southern Canadian Cordillera.* Atlas of Deep-Water Outcrops: American Association of Petroleum Geologists, Studies in Geology, 2007. **56**.

[46]. Armitage, D.A., et al., *The influence of mass-transport-deposit surface topography on the evolution of turbidite architecture: The Sierra Contreras, Tres Pasos Formation (Cretaceous), southern Chile.* Journal of Sedimentary Research, 2009. **79**(5): p. 287-301.

[47]. Di Celma, C.N., et al., *Spatial and temporal evolution of a Permian submarine slope channel–levee system, Karoo Basin, South Africa.* Journal of Sedimentary Research, 2011. **81**(8): p. 579-599.

[48]. Noad, J.J., *The sedimentary evolution of the tertiary of eastern Sabah, Northern Borneo*. 1999, Birkbeck (University of London).

[49]. Hutchison, C. and T. Dhonau, *Deformation of an Alpine ultramafic association in Darvel bay, east Sabah, Malaysia.* Geol. Mijnbouw, 1969. **48**(5): p. 481-494.

[50]. Tongkul, F., *Structural style and tectonics of Western and Northern Sabah.* Geological Society of Malaysia, Bulletin, 1990.

A review of walking energy harvesting using piezoelectric materials

Elham Maghsoudi Nia[1, a], **Noor Amila Wan Abdullah Zawawi**[2,b*], **Balbir Singh Mahinder Singh**[3, c]

[1,2]Department of Civil and Environmental Engineering, Universiti Teknologi PETRONAS, 32610, Seri Iskandar, Malaysia
[3]Department of Fundamental and Applied Sciences, Universiti Teknologi PETRONAS, 32610, Seri Iskandar, Malaysia

*amilawa@utp.edu.my

Abstract. Harvesting kinetic energies is a sustainable method for generating electricity without depleting natural resources. The main mechanisms for kinetic energy harvesting are piezoelectric, electromagnetic, electrostatic or by using magnetostrictive materials. This study focuses on harvesting of walking energy and aims to compare different technologies used for converting of walking energy to electricity, and identify the most effective technology. Several types of harvester located on body of user to harvest kinetic energy of body during walking, while some pavement slabs are produced for harvesting energy. The paper concludes that the pavement equipped with the harvesters would be more reliable than the body located technologies since it is independent of the physiological parameters. Moreover, the piezoelectric transduction is more desirable due to its advantages such as simplicity and flexibility, while produce less current output than the electromagnetic transduction.

Keywords: kinetic energy harvesting, walking energy conversion, footsteps, piezoelectric generator, pavement

1. Introduction

Energy harvesting or scavenging is the process of capturing the wasted energy from naturally occurring energy sources, accumulating and storing it for later use [1]. In another definition by Kazmierski and Beeby [2] energy harvesting is the conversion of ambient energy, exist in the environment into electrical energy. Harvesting energy is one of the most promising techniques in response to the global energy problem without depleting natural resources. Energy harvesting typically refers to micro- to milli-watts small power generation systems that is developed as a method for replacing or augmenting batteries. It exploits kinetic, thermal, solar sources or electromagnetic radiation sources. The kinetic energy harvesting that is the purpose of this study, converts movement, mainly in form of vibrations, into electrical energy. An energy harvester has typically three main components; the micro-generator for converting ambient energy into electrical energy, the voltage booster to pumps up and regulate the generator voltage and the storage element [3]. Since vibration power generators are mainly resonant systems, the maximum power is generated when the resonant frequency of the generators matches ambient vibration frequency [4]. Regarding the scope of the study that is harvesting kinetic energy of

Content from this work may be used under the terms of the Creative Commons Attribution 3.0 licence. Any further distribution of this work must maintain attribution to the author(s) and the title of the work, journal citation and DOI.
Published under licence by IOP Publishing Ltd

walking, the frequency of the human movements is less than 10 Hz [5]. Harvesting kinetic energy of the human body including walking energy is one of the methods of providing electricity for low-powered devices through implementing energy harvesting technologies, which in the individual has the potential to become a power generator [6].

2. Principles of kinetic energy harvesting

There are different types of vibration energy to harvest, that have been studied by several scholars, such as human motion [7], ocean wave [8] or harvesting strain from beam elements in critical structures [9]. Energy harvesting from mechanical loading generated in the ground in the shape of compressive forces while people move across the floor, is a sustainable method to generate electrical energy [10]. The generic model of kinetic energy harvesters was first developed by Williams and Yatus [4, 11]. Kinetic energy harvesters, also known as vibration power generators, extract electrical power by employing one or a combination of different mechanisms including piezoelectric, electromagnetic, electrostatic or by using magnetostrictive materials [4]. Many scholars have studied the properties of theses transductions, the advantages and the disadvantages of each mechanism [1]. This study aims to focus on the harvesting of walking energy as a type of kinetic energy. Hence, firstly it is needed to study harvesting the kinetic energy of body during walking and then focuses on harvesting of the walking energy using piezoelectric harvesters in the floor.

3. Harvesting kinetic energy of body during walking

Study of energy harvesting has recently gained attention in order to power portable electronic devices [12, 13]. Human activities as energy sources [14] are considered for harvesting by employing electromagnetic, piezoelectric, or electrostatic transduction methods [12, 15, 16]. In order to maximize power output, the resonant frequency of a harvester (fr), should match the dominant frequency of motion (fm) [7]. Generally, motion energy availability increases as fm increases [17]. The idea of harvesting energy from human motion is based on the fact that the amount of energy used by the body per day, is 1.07×107 J [18], an amount equivalent to around 800 AA (2500 mAh) batteries [19].

Walking is the main motion in normal human lives, hence, is considerable vibration energy for harvesting [7]. Walking is an economical energy harvesting approaches [14]. [Romero, Galchev [20]] estimated that 1 mW/cm3 power could be produced from human walking that this level of power would energize low-power applications [15]. Therefore, several harvesters are under development and different ways of harvesting motion energy, including inertial energy harvesters, integrating a piezoelectric system, the technology of electrostatic generators based on electroactive polymers (EAPs), or electromagnetic energy harvester. Inertial forces have been used for energy generation either inside shoes [21], within the shoe sole [6, 22], [23] that harvest energy from footfalls or mobile phone chargers integrated in backpacks[24] or phones [25]. Gorlatova, Sarik [7] focus on inertial energy harvesters and studied human physical parameters when harvesting walking energy. They showed that the taller half of the participants, harvested more power than the shorter half (around 20%). Moreover, the average power for going downstairs is significantly higher than for going upstairs. As the measurements of the Gorlatova, Sarik [7] study demonstrate, for prolonged activities, average absolute deviation of the acceleration, frequency of motion, and the average power vary considerably over time due to the physiological parameters changes [7].

The heel-strike generators using piezoelectric and electrostatic devices [26] with few mW power outputs has gained attention due to the rather high impact forces generated during walking. The main advantage of piezoelectric materials that are mainly used for heel strike devices, is their simplicity for incorporating into a shoe. However, the piezoelectric elements as energy harvesters are proper for low frequencies conditions and the power output of this technology is not enough to power high-consumption devices [27] and limited to nearly 100 mW due to the small displacement and the high generated voltage [28]. The technology of EAPs (electrostatic generators based on electroactive polymers) as the best choice for heel-strike devices have a high power-to-weight ratio [19]. Furthermore, some researches show that use of electromagnetic can generate energy with higher power output [12, 13]. Romero,

IOP Conf. Series: Materials Science and Engineering **291** (2017) 012026 doi:10.1088/1757-899X/291/1/012026

Warrington [15] suggested considering careful design for a body located harvester used the available energy of the body to power an electronic device. To increase the power output at low vibration frequency, Zhang, Wang [29] developed the energy harvesters to convert low-frequency vibrations to a higher frequency by employing the frequency up-conversion technique [30, 31]. Their simulations of power generation through magnet and coil arrays show that the high energy-conversion efficiency can be maintained at low vibration frequency and large vibration amplitude [29]. As a consequence, according to Riemer and Shapiro [19] the most power output refers to the technology of EAPs for heel-strike devices. Table-1 shows the output power of different technologies located on the body.

Table 1. Comparison of different technologies located on the body for converting energy of body motion during walking to electrical energy.

Ref.	Type of Technology	Output Power
	Inertial energy harvesters:	
[23]	Body's own inertia (compression of the shoe sole)	0.8 W
[32, 33]	Inertial energy harvesters for a small wireless device, from walking	100–200 μW
[7]	Inertial energy harvesters fits for Internet of Things applications:	
	for trouser pocket placement	202 μW
	for waist belt placement	180 μW
	for shirt pocket sensing unit placement	155 μW
[24]	Inertial forces inside backpacks	Few milliwatts to a
[21]	Inertial forces inside backpacks Shoes	few watts
	Piezoelectric elements:	
[34]	Piezoelectric generator harvesting the energy of footfall during walking	150-675mW
[28]	The piezoelectric elements for low frequencies conditions and low-consumption devices	Nearly 100 mW
[35, 36]	A cymbal-shaped piezoelectric	53mW
[37]	Two back-to-back unimorphs	8.4mW
[22]	Harness the heel strike using a unimorph strip	1.8mW
[37]	The PVDF stack	1.3mW
[22]	A stack of Polyvinylidinefluoride (PVDF) sheets shaped	1.1mW
[38]	A piezoelectric vibration generator	375 μW
	The technology of EAPs:	
[19]	The technology of EAPs for heel-strike devices	0.8 W
	From the heel strike (for a walking person weighing 80 kg at around 4 km/h)	2 W
	The knees and the ankles (for a person weighing 80 kg)	34 W and 20 W, respectively
	Electromagnetic energy harvester:	
[24]	Spring-loaded energy harvesting backpack with a magnetic generator	7 W power
[14, 24]	The biomechanical energy harvester consists of an aluminum chassis and a generator located on a customized knee brace	(4.8±0.8) W
[29]	A hand-held electromagnetic energy harvester from body motion located in a backpack of a user at 3.58m/s walking speed	32 mW
[15]	Electromagnetic induction generator located alongside of the ankle for implantable devices	3.9 μW

4. Harvesting kinetic energy of walking using piezoelectric harvesters in the floor

The main technologies used for harvesting kinetic energy of walking through the floor are based on piezoelectric and electrical induction generators, and electrostatic generators based on electroactive polymers (EAPs) [19]. Generally, there are different methods for producing electrical energy from moving people through the pavement based power generation technologies. An unusual approach is the use of pressure changes in the ground when people walk, that is exposed to permanently fluctuating pressure amplitudes. For example, the movement of a dance floor can be used to generate electricity through employing an electromagnetic generator. However, in order to generate a considerable electric power, there is a need to a relatively large deflection of the floor, up to 10 mm. Moreover, it needs a complicated structure that impose high assembling costs. Another example is piezoelectric harvesters in the floor of a subway ticket machine. The harvesters consist of piezoceramic and there is no need to complicated mechanical structures [10]. The rubber mats containing piezoelectric sensors were integrated along ticket gates for Tokyo subways in 2008 [6]. Since in this method, the energy conversion is based on the piezoelectric effect, there is no need for deflection of the ground. The efficiency of this energy conversion depends to the thickness and properties of the piezoelectric material, and the force affecting the piezoelectric material. Higher thicknesses and higher forces generate higher amount of surface charges. Among the different piezoelectric materials, only Polyvinylidene fluoride (PVDF) offers a long lasting functionality under these conditions at a comparable low price. PVDF is incombustible, chemically resistant and can be processed by injection molding. It is less expensive compared to lead zirconate titanate (PZT), and is unbreakable under harsh conditions. Nevertheless, the pure piezoelectric coefficients, the relatively low Curie-Temperature (105°C) and the necessity to stretch and to polarize the material are disadvantages of PVDF [10].

Bischur and Schwesinger [10] studied piezoelectric PVDF-foils for converting dynamic compressible mechanical loads into electrical energy. They found that the higher remnant polarization causes the better the energy conversion. Moreover, electricity generated rises with an increasing number of windings. They concluded that piezoelectric PVDF-foils appear to be very promising compared to PZT ceramic. PVDF as a polymer material is flexible and resistant in mechanical destruction forces. While, PZT as an inorganic ceramic is brittle and needs protection in a harsh environment [10]. Another example of walking energy harvesting through the floor is implementation of both technologies of piezoelectric and electromagnetic along pedestrians footpaths for Crowd Farm project by MIT in 2007 [6].

Wu, Tsung-Tsi, et al. Wu, Wang [39]studied a harvesting floor designed with piezoelectric material, which can convert extra energy of walking motion into electrical energy. Under low frequency condition, the energy generation from piezoelectric material has a low efficiency that is insufficient to drive a wireless module. In order to elevate the energy power of piezoelectric, they adopted the plucked method in their research, which the generated energy is approximately 10 times more than the energy of the forced method. This method changes the frequency from lower one like footstep to higher one like resonant frequency of the beam. Due to the small size of the piezoelectric material, and high power density, the size of the harvesting floor can be minimized [39]. In addition, [Kumar and Chaturvedi [40]] simulated and experimented a model of energy harvester floor tile and concluded that output voltage would be increased by employing high coupling coefficient piezoelectric material, increasing the number of piezoelectric diaphragm per unit area and parallel arrangement of them.

Furthermore, there are some achievements and studies on harvesting footstep through implementing different technologies in order to generate power for lighting LED lamps. One of the products that harvest kinetic energy of people stepping on them and immediately convey tiny bursts of electricity to nearby appliances, is the recycled rubber "Pavegen" paving slabs [41]. The Pavegen tile is a promising method of harvesting footsteps energy and is designed to compress five millimetres when someone steps on them. Since Pavegen has not share the precise mechanism of converting harvested kinetic energy into electricity, some guess that it utilizes piezoelectric sensors [42], while others believe using of electromagnetic technology. The efficiency of tiles in power generation is rather high, which each tile harvests 6 to 8 joules of energy per footstep. Each step produces only enough electricity to keep an LED-powered street lamp lit for 30 seconds. The system uses 5 % of the power to light the central lamp and the other 95 % of the power either can be stored in internal or external batteries for up to three days or used directly for external applications. Hence, when no central light is required, the output is 100 % of the power [6]. In their first commercial application, they expected that by scattering 20 tiles along the central crossing between London's Olympic stadium and Westfield Stratford City mall, and estimating 30 million customers, can power half of outdoor lighting at vast London shopping mall. Kemball-Cook, the founder and CEO of Pavegen, mentioned another example of using this slab recently in a big outdoor festival received over 250,000 footsteps that was enough to charge 10,000 mobile phones. He added that it is also easy to install the Pavegen slab as a retrofit on existing pavements, can match their exact dimensions and be replaced one slab with another. He also foresees Pavegen systems to power off-grid appliances such as public lighting, illuminated street maps and advertising, and can be used in dense human traffics areas [41]. The main issue of green slabs of Pavegen is its price. However, the company hopes to see the price drop to below $100 per tile in the near future.

Additionally, Vanz and Karakiewicz [6] proposed a power generating device for safety lighting along the specific stairways in Melbourne, and focused on ergonomics, comfort and safety of the pedestrian as well. They considered four conditions including force of footstep, technology, service provided, and context in their study. They measured the selected ten piezoelectric sensors, generating 300 μJ per actuation (EG300), per tread and two steps, and placed 20 LEDs along two steps. Since per sensor can light the 20 LEDs, hence, they would be powered for 1.5 hour per day. By increasing the number of sensors per tread up to 15 and distribution of 20 LEDs along three stair treads, which is sufficient for safety purpose the duration of the lighting increases to over 3.5 hours. It can be estimated that by considering 10 sensors (EG300) along one stair tread, the power generating system can keep 200 LED light for 10 hours at night time for a higher pedestrian's traffic pathway, such as the subway in Tokyo, with around 4 million pedestrians daily. By comparing the developed device based on the piezoelectric sensors with Pavegen technology, it could be concluded that the proposed prototype is more efficient, due to the physical occupied space by the systems, and the system efficiency in capturing the footstep energy using the flexibility of the prototype design [6].

As a consequence, both piezoelectric harvesters located on the body and in the floor, can generate power from a microwatt to few watts. However, harvesters located on the body are dependent on the location of the generator and average of the acceleration, frequency of motion, and the power vary, significantly. Therefore, to gain a consistent output, implementing the harvester within the pavement slab is a better choice.

5. Conclusion

The study has reviewed the different methods of harvesting footsteps, the power output and in some cases the applications. By comparing the harvesting energy technologies located on the body and pavement, it is revealed that for body located harvester the power output depends on the physiological parameters. Therefore, it is better to implement the harvester within the pavement slab to gain a consistent output. Furthermore, in comparison between the different types of transductions used in the equipped pavements, although, the output current of the electromagnetic mechanism is high, the piezoelectric technology with the simple structure, flexibility of the design and geometry, small size of the sensors, the ability of easily meshing into hybrid materials has made the piezoelectric transduction desirable. Moreover, four proposed interrelated conditions, play a main role in achieving the desirable results. It is obvious that the number of pedestrian is the main factor when selecting the type of technology and can be effective on the generated power output. In the other word, the type of transduction has to be selected according to the number of footsteps and the needed services. The study recommends more exploration on the piezoelectric properties to optimize the power output in order to use the technology for the pavement in different application.

6. References

[1] Andriopoulou, S., A review on energy harvesting from roads. 2012.
[2] Kazmierski, T.J. and S.P. Beeby, Energy harvesting systems. 2010: Springer.
[3] Kazmierski, T.J. and L. wang, Modeling, Performing Optimisation and Automated Design of Mixed-Technology Energy Harvester System, in Energy harvesting systems, T.J. Kazmierski and S.P. Beeby, Editors. 2010, Springer.
[4] Zhu, D. and S. Beeby, Kinetic Energy Harvesting, in Energy harvesting systems, T.J. Kazmierski and S.P. Beeby, Editors. 2010, Springer.
[5] Roundy, S., P.K. Wright, and J. Rabaey, A study of low level vibrations as a power source for wireless sensor nodes. Computer communications, 2003. 26(11): p. 1131-1144.
[6] Vanz, E. and J. Karakiewicz, Pedestrian as Generator: Implementing a Stand-Alone Piezo Power Generating Device in the Urban Context, in Computer-Aided Architectural Design Futures. The Next City-New Technologies and the Future of the Built Environment. 2015, Springer. p. 154-171.
[7] Gorlatova, M., et al. Movers and shakers: Kinetic energy harvesting for the internet of things. in The 2014 ACM international conference on Measurement and modeling of computer systems. 2014. ACM.
[8] Lewis, A., et al., Ocean Energy, in In IPCC Special Report on Renewable Energy Sources and Climate Change Mitigation [O. Edenhofer, R. Pichs-Madruga, Y. Sokona, K. Seyboth, P. Matschoss, S. Kadner, T. Zwickel, P. Eickemeier, G. Hansen, S. Schlömer, C. von Stechow (eds)] 2011: Cambridge, United Kingdom and New York, NY, USA.
[9] Pfeifer, K.B., S. Leming, and A. Rumpf, Embedded self-powered micro sensors for monitoring the surety of critical buildings and infrastructures. SAND2001-3619 Unlimited Release Printed November, 2001. 3: p. 3q6.
[10] Bischur, E. and N. Schwesinger. Energy harvesting from floor using organic piezoelectric modules. in Power Engineering and Automation Conference (PEAM), 2012 IEEE. 2012. IEEE.
[11] Williams, C. and R.B. Yates, Analysis of a micro-electric generator for microsystems. sensors and actuators A: Physical, 1996. 52(1): p. 8-11.
[12] Romero, E., R. Warrington, and M. Neuman, Energy scavenging sources for biomedical sensors. Physiological measurement, 2009. 30(9): p. R35.
[13] Beeby, S.P., M.J. Tudor, and N. White, Energy harvesting vibration sources for microsystems applications. Measurement science and technology, 2006. 17(12): p. R175.
[14] Donelan, J.M., V. Naing, and Q. Li. Biomechanical energy harvesting. in Radio and Wireless Symposium, 2009. RWS'09. IEEE. 2009. IEEE.
[15] Romero, E., R.O. Warrington, and M.R. Neuman. Body motion for powering biomedical devices. in Engineering in Medicine and Biology Society, 2009. EMBC 2009. Annual International Conference of the IEEE. 2009. IEEE.
[16] Starner, T., Human-powered wearable computing. IBM systems Journal, 1996. 35(3.4): p. 618-629.
[17] Mitcheson, P.D., et al., Energy harvesting from human and machine motion for wireless electronic devices. Proceedings of the IEEE, 2008. 96(9): p. 1457-1486.
[18] Exercise, S.A., Disease Prevention: in Exercise Physiology: Energy, Nutrition and Human Performance: McArdle, WD, Katch, FI, Katch, VL. 2001, Lippincott, Williams and Wilkins.
[19] Riemer, R. and A. Shapiro, Biomechanical energy harvesting from human motion: theory, state of the art, design guidelines, and future directions. Journal of neuroengineering and rehabilitation, 2011. 8(1): p. 22.
[20] Romero, E., et al. Micro energy scavengers. in Electronic Proc. COMS. 2008.
[21] Duffy, M. and D. Carroll. Electromagnetic generators for power harvesting. in Power Electronics Specialists Conference, 2004. PESC 04. 2004 IEEE 35th Annual. 2004. IEEE.
[22] Kymissis, J., et al. Parasitic power harvesting in shoes. in Wearable Computers, 1998. Digest of

IOP Conf. Series: Materials Science and Engineering **291** (2017) 012026 doi:10.1088/1757-899X/291/1/012026

Papers. Second International Symposium on. 1998. IEEE.

[23] Kornbluh, R.D., et al. Electroelastomers: applications of dielectric elastomer transducers for actuation, generation, and smart structures. in SPIE's 9th Annual International Symposium on Smart Structures and Materials. 2002. International Society for Optics and Photonics.

[24] Rome, L.C., et al., Generating electricity while walking with loads. Science, 2005. 309(5741): p. 1725-1728.

[25] Lin, G., et al., Harnessing power through electromagnetic induction utilizing printed coils. 2012, Google Patents.

[26] Starner, T. and J.A. Paradiso, Human generated power for mobile electronics. Low power electronics design, 2004. 45: p. 1-35.

[27] Almouahed, S., et al., The use of piezoceramics as electrical energy harvesters within instrumented knee implant during walking. Mechatronics, IEEE/ASME Transactions on, 2011. 16(5): p. 799-807.

[28] Niu, P., et al. Evaluation of motions and actuation methods for biomechanical energy harvesting. in Power Electronics Specialists Conference, 2004. PESC 04. 2004 IEEE 35th Annual. 2004. IEEE.

[29] Zhang, Q., Y. Wang, and E.S. Kim, Power generation from human body motion through magnet and coil arrays with magnetic spring. Journal of Applied Physics, 2014. 115(6): p. 064908.

[30] Sari, I., T. Balkan, and H. Kulah, An electromagnetic micro power generator for low-frequency environmental vibrations based on the frequency upconversion technique. Microelectromechanical Systems, Journal of, 2010. 19(1): p. 14-27.

[31] Kulah, H. and K. Najafi. An electromagnetic micro power generator for low-frequency environmental vibrations. in Micro Electro Mechanical Systems, 2004. 17th IEEE International Conference on.(MEMS). 2004. IEEE.

[32] Huang, H., G.V. Merrett, and N.M. White, Human-powered inertial energy harvesters: the effect of orientation, location and activity on obtainable power. Procedia Engineering, 2011. 25: p. 815-818.

[33] Von Büren, T., et al., Optimization of inertial micropower generators for human walking motion. Sensors Journal, IEEE, 2006. 6(1): p. 28-38.

[34] Antaki, J.F., et al., A gait-powered autologous battery charging system for artificial organs. ASAIO journal, 1995. 41(3): p. M588-M595.

[35] Kim, H., et al., Consideration of impedance matching techniques for efficient piezoelectric energy harvesting. Ultrasonics, Ferroelectrics, and Frequency Control, IEEE Transactions on, 2007. 54(9): p. 1851-1859.

[36] Kim, H.W., et al., Energy harvesting using a piezoelectric “cymbal” transducer in dynamic environment. Japanese Journal of Applied Physics, 2004. 43(9R): p. 6178.

[37] Shenck, N.S. and J.A. Paradiso, Energy scavenging with shoe-mounted piezoelectrics. IEEE micro, 2001(3): p. 30-42.

[38] Roundy, S. and P.K. Wright, A piezoelectric vibration based generator for wireless electronics. Smart materials and Structures, 2004. 13(5): p. 1131.

[39] Wu, T.-T., et al. Analysis of high efficiency piezoelectric floor on intelligent buildings. in SICE Annual Conference 2010, Proceedings of. 2010. IEEE.

[40] Kumar, D. and P. Chaturvedi, Piezoelectric Energy Harvesting from Vibration Induced Deformation of Floor Tiles, in Sixth National Power Electronics Conference (NPEC). 2013: IIT Kanpur, India.

[41] Webster, G., Green sidewalk makes electricity--one footstep at a time. 2011, CNN.

[42] Fujimoto, A., Energy Harvesting Flooring. 2014, tanford University.